The

LITTLE
ADSORPTION
BOOK

A Practical Guide for Engineers and Scientists

The

LITTLE
ADSORPTION
BOOK

A Practical Guide for Engineers and Scientists

Diran Basmadjian

Department of Chemical Engineering and Applied Chemistry
University of Toronto
Toronto, Ontario, Canada

CRC Press
Boca Raton   New York   London   Tokyo

Acquiring Editor:	Norm Stanton
Project Editor:	Andrea Demby
Marketing Manager:	Susie Carlisle
Direct Marketing Manager:	Becky McEldowney
Cover design:	Dawn Boyd
PrePress:	Gary Bennett
Manufacturing:	Sheri Schwartz

Library of Congress Cataloging-in-Publication Data

Basmadjian, Diran.
    The little adsorption book : a practical guide for engineers and
scientists / Diran Basmadjian.
       p.   cm.
    Includes bibliographical references and index.
    ISBN 0-8493-2692-3 (alk. paper)
    1. Adsorption. I. Title.
TP156.A35B37 1996
660′.293—dc20

             96-32834
             CIP

No claim to original U.S. Government works
International Standard Book Number 0-8493-2692-3
Library of Congress Card Number 96-32834
Printed in the United States of America 1 2 3 4 5 6 7 8 9 0
Printed on acid-free paper

PREFACE

This book was prompted by the numerous enquiries the author has received over the years from students, both undergraduate and graduate, industrial practitioners, and others who, having known adsorption only as a dark and forbidding subject, were looking for answers to immediate, often mundane but pressing, problems. These problems frequently involve no more than a preliminary sizing of a unit, estimation of bed and purge requirements, or alleviating the malfunction of an existing unit. In the case of natural systems, one wishes to know the degree of contamination of soils and river beds or, more importantly, the expected time of recovery.

These questions, which one needs to answer for a first assessment, or indeed a first grasp, of adsorption processes, are not addressed in textbooks and can only be extracted with difficulty from the general literature. They are shunned in the traditional adsorption texts, or indeed in Perry's *Handbook for Chemical Engineers*.

We have attempted to fill the gap by drawing, in the first instance, on algebraic mass and energy balances, which arise from equilibrium models ("equilibrium theory"). They are liberally supplemented with simple "operating diagrams" and together these two tools, supplemented by certain rules, provide lower bounds to important system parameters, such as on-stream time and sorbent or purge requirements. Single-component systems of various types, as well as more complex ones involving multicomponent, adiabatic, and chromatographic operations, are addressed. Equilibria are considered mainly in the context of Henry constants, including those that pertain to water-soil systems. Finally, we graft onto these results the effects of transport resistance to complete the picture. We do not touch on PSA (pressure saving adsorption), which is well covered elsewhere.

The book is primarily addressed to students as well as physicists, chemists, and engineers who have to deal with adsorption phenomena. The simple style should make it easy reading even to the uninitiated environmentalists (on whose work we have drawn heavily), who will benefit from the new perspective presented in this volume. Its relatively short length makes it particularly useful for workshops and for introductory courses at the undergraduate and graduate levels.

Throughout this slim volume the theme has been to keep the treatment simple and understandable, but it is not our intention to exclude academic use, having in mind particularly those individuals who struggle to make a first elementary presentation of adsorption processes.

The author owes much to the help given by Dr. Douglas M. Ruthven, University of Maine, and Dr. Kent S. Knaebel, Adsorption Research, Inc. Their advice and support have been invaluable. Dr. Donald Mackay, Trent University, introduced the author to the methodology used by environmentalists, thus enabling him to draw on their work. A good deal of what appears in the chapter on adiabatic sorption is based on the seminal work of Dr. C. Y. Pan, Alberta Research Council.

It remains for me to thank my collaborator and fellow-Newfoundlander, Arlene Fillatre, for her devoted efforts to bring this work to a successful conclusion.

Diran Basmadjian, Professor

Department of Chemical Engineering and Applied Chemistry
University of Toronto
Toronto, Ontario, Canada

TABLE OF CONTENTS

1.   Introduction 1

2.   Equilibrium Effects: Single-Component Isothermal
     Systems 7
     2.1   Equilibrium Isotherms 7
     Illustration... 8
     2.2   Simple Mass Balances for Type I Isotherms 9
     Illustration... 14
     2.3   General Column Operation 15
     2.4   Column Sorption for Type I Isotherms................. 16
     Illustrations 19
     2.5   Column Sorption for Type III Isotherms 20
     2.6   Column Sorption for Linear Isotherms 21
     Illustrations 23
     2.7   Column Sorption for Inflecting Isotherms 25
     Illustrations 29

3.   Equilibrium Effects - The Binary Langmuir Case 33
     3.1   Binary Adsorption and Desorption 35
     3.2   Creative Doodling: Prediction of Various Binary
           Adsorption Profiles 37
     3.3   Mutual Displacement of Individual
           Components 40
     Illustrations 43

4.   Equilibrium Effects: Adiabatic Sorption 53
     4.1   Magnitude of the Temperature Rise ΔT
           Qualitative... 54
     4.2   Interaction of Temperature and Concentration
           Fronts 55
     4.3   Representation in the q-Y Diagram 59
     4.4   Design Equations, Minimum Bed and Purge
           Requirements 62

4.5   Maximum Enrichment Attainable in Hot Purge
         Desorption; Steam Regeneration............................ 63
      Illustrations 65

5.   Equilibrium Effects: Multicomponent Sorption 75
      5.1   Linear Isothermal Systems 76
      Illustration.. 76
      5.2   Nonlinear Isothermal Systems 77
      Illustrations 78

6.   Equilibrium Effects: Linear Chromatography............... 81
      6.1   General Features.. 81
      Illustrations 83

7.   Adsorption Equilibria.. 87
      7.1   Henry Constants... 87
      7.2   Sorption Saturation Capacities............................ 91
      Illustration.. 93

8.   Sorption with Transport Resistance 95
      8.1   General Features.. 95
      8.2   The Separation Factor r... 98
      8.3   Transport Coefficients 99
      8.4   The Design Charts 101
      8.5   Conditions for Preventing Instantaneous
            Breakthrough 110
      8.6   Grafting of Transport Effects onto Equilibrium
            Theory (No Transport Resistance) 110
      8.7   Conditions for the Attainment of Sharp
            Sorption Fronts 111
      Illustrations 112

Nomenclature 127

Bibliography... 129

Figure Titles... 131

Index 135

Chapter 1

INTRODUCTION

We start this book in an unconventional way by first considering the response of a mass of sorptive material to a percolating fluid containing sorbable solutes, or devoid of them. This is the problem which the novice, or the experienced engineer, wishes to address above all others. We leave aside for the time being consideration of equilibria and interphase transport, although their qualitative impact will be felt almost immediately. We make a point of separating the two, addressing first the role of equilibrium on the dynamics of sorption and grafting onto these results the effect of transport resistance. This is no more than good chemical engineering practice (see the concepts of equilibrium stage and stage efficiency), but the novelty of a stationary phase is often seen as a hindrance which has led to a general spurning of this approach. This need not be and should not be. Equilibrium effects can, in many instances, be quickly calculated and provide precisely the kind of information one needs for preliminary performance estimates. Although this approach is not explicitly covered elsewhere, much useful background information can be gleaned from several authoritative monographs.[1-5]

A third departure from convention shows in our tendency often to content ourselves with approximate answers, or bracketing the true answer. Given the usual time constraints, as well as uncertainty in equilibrium and transport parameters (or worse: a total absence of them), this is no more than what common sense dictates.

Finally, we make a pledge to avoid differential equations. We shall be guided by a paraphrased version of a saying by Canadian Prime Minister Mackenzie King: PDEs if necessary, but not necessarily PDEs. It has been our experience that complex expressions can often be avoided by what we call "creative doodling", i.e., the use of graphical constructions. These are not only easy to perform, but are oftentimes more eloquent than precise numerical or analytical expressions. We do use analytical expressions, but these are

1

all algebraic in form and surprisingly few in number. Although simple in appearance and context — they are essentially mass and energy balances — the expressions derive from a more profound basis, i.e., the constitutive partial differential equations of sorption processes and their companion theory of characteristics. We outline some simple derivations of the mass balances in Section 2.2 and proceed to apply the results directly to problems of interest. When necessary, results of the underlying theory are cast into "rules" which establish the trajectories and pathways of sorption processes. Together this trio of tools — graphical constructions, algebraic balances, and the trajectory rules — provide a powerful and rapid means for obtaining lower bound answers to a host of complex problems. The lower bound restriction is subsequently removed by incorporating the effect of transport resistance to complete the circle. A good deal of the background material was laid down in previous publications by the author.[6-9]

Although similar results can be obtained by computer simulations of the full differential equations model, the present approach has the advantage of much greater rapidity and of providing deeper insight into the physics of sorption processes. The separation into equilibrium and kinetic effects, in particular, allows us to adjust relevant parameters quickly to analyze or improve performance, or pinpoint the source of difficulties.

We conclude this section with a number of items designed to convey basic properties of commercial sorbents and sorption processes.

In Figure 1.1 we have sketched the principal features of a classical adsorption process along a type I (or Langmuir-type) isotherm. The sorption front stabilizes to an S-shaped, or "constant pattern" form, which results in a division into sections of saturated and clean bed, separated by a mass transfer zone represented by the S-shaped curve. Under equilibrium conditions the mass transfer zone disappears and the sorption front propagates as a rectangular discontinuity. Concentration variations as a function of distance z are referred to as profiles (left side of diagram), while variations in time t at a fixed position yield the concentration histories, often referred to as breakthrough curves. The concentration units chosen, kilograms solute per kilogram sorbent or carrier, deviate somewhat from the traditional molar concentration units or partial pressures, but carry the advantage of yielding particularly simple expressions of direct practical interest, such as bed and purge requirements and facilitating the use of graphical constructions.

In Figure 1.2 we display the various types of equilibrium isotherms encountered in sorption operations and follow this up in Figure 1.3 with a demonstration of their occurrence in a particular practical application, that of moisture uptake on commercial desiccants. Both convex and concave curves make their appearance, as well as isotherms with inflections. These various types of equilibrium curves are addressed more thoroughly in subsequent sections.

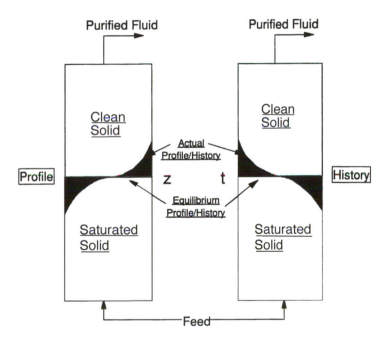

FIGURE 1.1

Typical concentration front obtained during adsorption. The front has stabilized to a "constant pattern" form. Profiles and histories ("breakthrough curves") are mirror images of each other. G_b, carrier mass velocity, kg m^{-2} s^{-1}, $= \rho_f v$; v, superficial fluid velocity, m s^{-1}; V, propagation velocity of solute front, m s^{-1}; ρ_b, bed density, kg m^{-3}; ρ_f, fluid density, kg m^{-3}; Y, fluid phase concentration, kg solute/kg carrier; q, solid phase concentration, kg solute/kg solid.

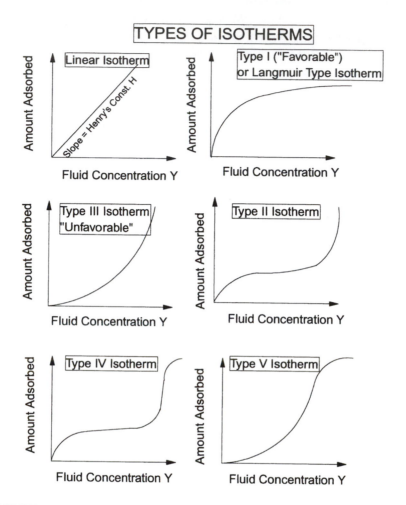

FIGURE 1.2
The six types of equilibrium isotherms encountered in sorption operations.
Types II, IV, and V inflect. Type I is of the classical Langmuir form. At low
concentrations, all isotherms converge to the linear form shown in the upper
left-hand diagram.

Table 1.1 summarizes properties of the most important commercial adsorbents, while Table 1.2 is meant to convey a sense
of the range of parameter values encountered in sorption operations. These tables provide lower and upper bounds of physical
properties and dimensions likely to be encountered in sorption
operations and are designed to aid in estimating operational
parameters.

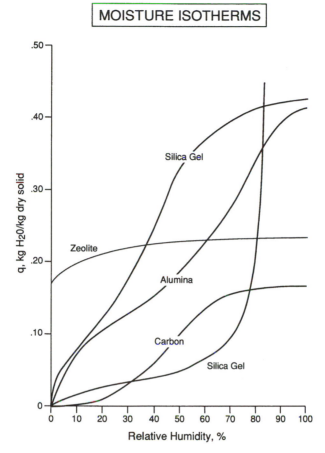

FIGURE 1.3
Moisture uptake on various commercial sorbents. Sorption on zeolites shows a type I form and is most effective at low humidities. Silica gel and alumina have high uptakes at high humidities. Carbon has a low affinity for moisture.

TABLE 1.1
Properties of Commercial Adsorbents

| | Densities, g/cm³ | | Diameter | | |
| | | | Particle, | | Surface |
Type	Bulk	Particle	mm	Pore, Å	Area, m²/g
Activated carbon	0.44–0.48	0.75–0.85	1–5	15–20	950–1250
Activated alumina	0.60–0.85	1.2–1.4	2–12	25–50	250–350
Silica gel	0.40–0.75	1.2	1–7	20–140	350–700
Zeolites	0.60–0.70	1.0–1.7	1–5	4–10	—

TABLE 1.2

Range of Fixed Bed Parameters

			Superficial Velocity, m/s	
	Diameter, m	Packed Height, m	Gas	Liquid
Laboratory	$5 \cdot 10^{-2}$–10^{-1}	10^{-2}–1	10^{-2}–1	10^{-4}–10^{-3}
Plant	0.5–3	0.5–5	0.5–5	10^{-2}–10^{-1}

Illustration 1.1: Drying of Gases

A choice of adsorbent is to be made for drying a gas to low relative humidities prior to low temperature treatment or cryogenic liquefaction (see Figure 1.3).

- Activated carbon, being a hydrophobic substance, has low capacities, particularly at low relative humidities and is thus ruled out as a desiccant.
- Activated alumina (Al_2O_3) and silica gel (SiO_2) have high capacities at high humidities, but only modest uptakes at low moisture levels.
- Zeolites have the highest uptakes at low humidity and moderate capacities at high humidities.

Zeolites recommend themselves particularly for "deep drying", i.e., the attainment of low dew points required in cryogenic liquefaction operations. Silica gel and alumina are suitable for more modest dew point requirements. Use has also been made of composite beds made up of layers of Al_2O_3/zeolite or SiO_2/zeolite. The alumina and silica gel layers serve to remove the bulk of high humidity moisture by virtue of their high capacities in that region, the zeolite playing the role of final scavenger of moisture. The greater efficiency of this mode is somewhat offset by the greater operational complexity of composite beds.

We note that free moisture has to be removed prior to introduction of the gas to the bed to guard against damage to the adsorbent. Silica gel fractures on contact with liquid water and both alumina and zeolites undergo structural deterioration. A guard bed of inert particles and other appropriate filters or prior removal in a heat exchange device suggest themselves as remedies.

Chapter 2

EQUILIBRIUM EFFECTS: SINGLE-COMPONENT ISOTHERMAL SYSTEMS

2.1 EQUILIBRIUM ISOTHERMS

One component adsorption equilibria are traditionally divided into five categories: type I ("favorable") and type III ("unfavorable") are concave downward and upward, respectively, while the remaining ones — types II, IV, and V — are of an inflecting type. The latter category is associated with multilayer adsorption and capillary condensation. It arises in moisture uptake by sorbents and by natural products (wool, grain), and is generally seen in the sorption of vapors near their boiling point (Figure 1.2). It also makes its appearance in the form of effective equilibrium curves in adiabatic sorption.

Linear isotherms are usually identified by their slope, which equals the Henry constant H. They evolve from all other isotherms at sufficiently low concentrations and, conversely, revert to one of the types I to V at higher concentrations. Linear isotherms, in other words, do not stay linear over an unlimited concentration range. They are, nevertheless, an important subcategory with their own special properties, and are used extensively in modeling sorption of trace solutes and pollutants, or whenever the operational range of a process lies exclusively in the linear region. A discussion of this special case follows the sections devoted to type I and type III isotherms.

The celebrated Langmuir equation represents a special case of the type I isotherm and includes the Henry constant H to represent adsorption in the limit of vanishingly small concentrations. We cast it in the form

$$q = \frac{HY}{1+b'Y} \tag{2.1}$$

i.e., we retain our previous mass ratio concentration units in preference over partial pressures and molar concentrations. The reason, as previously noted, lies in the simplifications which result in formulating mass balances and limiting consumptions.

The following illustration demonstrates the conversion of conventional versions of the Langmuir isotherm to the form given by Equation (2.1).

Illustration 2.1: Conversion of Langmuir Forms to Mass Ratio Units

Given the classical Langmuir form

$$q'\left(\frac{mol}{g}\right) = \frac{ap}{1+bp} \quad \text{where } p = \text{partial pressure}$$

To convert to mass ratios, we first note that, according to the laws of partial pressures

$$\frac{p}{P_{Tot}-P} = \frac{M_2}{M_1}Y$$

where $M_{1,2}$ = molar mass of solute and carrier gas.

Solving for p we obtain

$$P = \frac{P_{Tot}(M_2/M_1)Y}{1+(M_2/M_1)Y}$$

Substituting into the original Langmuir equation and converting q' to mass units we obtain

$$q\left(\frac{kg}{kg}\right) = \frac{P_{Tot}M_2aY}{1+(M_2/M_1+bP_{Tot}M_2/M_1)Y}$$

where a and b are the original Langmuir constants.

In most practical situations, $p \ll P_{Tot}$ so that the expression reduces to the simple form

$$q\left(\frac{kg}{kg}\right) = \frac{P_{Tot}M_2aY}{1+\left(bP_{Tot}M_2/M_1\right)Y}$$

The penalty we pay is that the isotherm is dependent on total pressure. That quantity, however, rears its head sooner or later when the necessary conversions to mass units required in industrial practice have to be made.

For liquid systems, the conversion is considerably simpler. Here we have the original form

$$q'\left(\frac{mol}{g}\right) = \frac{aC}{1+bC} \quad \text{where } C = \text{concentration in } \frac{mol}{ml}$$

Straightforward conversion of C into mass ratios leads to the expression

$$q\left(\frac{kg}{kg}\right) = \frac{\rho \cdot a \cdot Y}{1+b\left(\rho/M_1\right)Y}$$

where a and b are again the original Langmuir constants and ρ is the solution density in kg/l assumed to be equal to the solvent density (valid for dilute solutions).

2.2   SIMPLE MASS BALANCES FOR TYPE I ISOTHERMS

Mass balances for sorption operations generally lead to partial differential equations in the dependent variables q and Y (solid and fluid phase concentrations) and the independent variables time t and distance z. We do not reproduce them here but instead make use of two fundamental consequences which arise from them. These are graphically depicted in Figure 2.1.

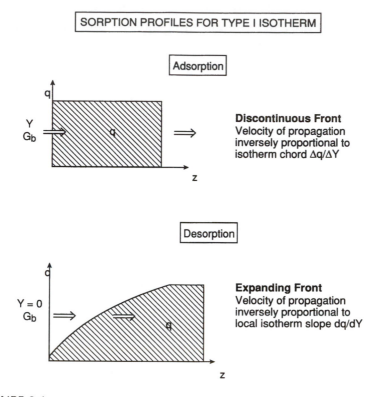

FIGURE 2.1
Propagation of discontinuous and expanding fronts arising in the sorption along a type I isotherm.

1. For adsorption along a type I isotherm, the sorption front penetrating the bed takes the form of a rectangular discontinuity which moves with a velocity inversely proportional to $\Delta q/\Delta Y$, i.e., the chord which connects the feed concentration with the initial bed condition on a q-Y diagram.

2. For desorption along a type I isotherm the sorption front leaving the bed has a continually expanding form, each point of which moves with a velocity inversely proportional to the local isotherm slope dq/dY.

This contrast in adsorption and desorption fronts profoundly affects sorption operations and is the preeminent reason why adsorption proceeds relatively fast, while desorption often leads to a long-drawn-out process. Kinetic effects (slow diffusion into the sorbent particle) may retard adsorption but the principal difference between the two operations is maintained. Other isotherm forms, which are taken up in more detail in subsequent sections, may give rise to combinations of sharp and elongated fronts. In the intriguing case of a type III isotherm, a complete reversal occurs: adsorption follows an elongated path, while desorption leads to a sharp discontinuous front. These features can all be accounted for by the fundamentals established here and in subsequent sections.

We can use the principles established above to derive simple mass balance equations, at least for the adsorption step. This is done by equating the cumulative amount of solute introduced with the accumulated amount within the bed. We assume for simplicity that the bed is initially clean. The procedure for a preloaded bed follows a similar pattern. Thus,

$$\text{Amount introduced to time } t = Y \cdot G_b \cdot A \cdot t$$
$$\text{Amount retained up to position } z = q \cdot \rho_b \cdot A \cdot z$$

From this one obtains the cumulative mass balance

$$q\, \rho_b\, z = Y\, G_b \cdot t \tag{2.2}$$

This is conveniently recast into the general form

$$V = z/t = \frac{G_b}{\rho_b \cdot \Delta q/\Delta Y} = \frac{\rho_f v}{\rho_b\, \Delta q/\Delta Y} \tag{2.3}$$

where z/t is recognized as the propagation velocity V of the discontinuous front which is inversely proportional to the chord $\Delta q/\Delta Y$ as stated before.

Similar but more elaborate procedures can be used to derive the corresponding expression for elongated desorption fronts, where the isotherm slope dq/dY now takes the place of the chord $\Delta q/\Delta Y$:

$$V = z/t = \frac{G_b}{\rho_b \cdot dq/dY} = \frac{\rho_f \cdot v}{\rho_b \cdot dq/dY} \tag{2.4}$$

We shall, in subsequent sections, refer to this expression as the design equation, since it is frequently used to establish certain design and operational parameters of the sorbent bed.

2.2.1 Specific Consumptions

Equations (2.3) and (2.4) can be conveniently recast to derive simple expressions for bed and purge requirements per unit of fluid treated. Starting with Equation (2.3)

$$z/t = \frac{\rho_f}{\rho_b} \frac{v}{\Delta q/\Delta Y}$$

we cross multiply and divide and multiply by the bed cross-sectional area A. Thus,

$$\frac{Az\rho_b}{Av\rho_f t} = \frac{1}{\Delta q/\Delta Y} \tag{2.5}$$

or

$$\frac{\text{Bed weight}}{\text{Weight of carrier fluid treated}} = \frac{1}{\Delta q/\Delta Y} = \frac{\Delta Y}{\Delta q} \tag{2.6}$$

This is a minimum bed requirement, since we are at this stage limiting ourselves to equilibrium conditions. It is, nevertheless, a useful expression which bears a simple relation to the chord connecting feed and initial bed conditions

$$\frac{\Delta q}{\Delta Y} = \frac{q_F - q_1}{Y_F - Y_1}$$

A similar relation can be derived for purge consumption. We start with Equation (2.4) and set dq/dY equal to the Henry constant H (= slope of the isotherm at the origin), i.e., we consider total desorption of the bed by purge. Proceeding as in the previous case we obtain

$$\frac{Az\rho_b}{Av\rho_f t} = \frac{1}{H} \tag{2.7}$$

Inverting the fractions this yields the following quantity

$$\frac{\text{Weight of purge required}}{\text{Bed weight}} = H \tag{2.8}$$

This is again a very simple, albeit limiting, expression which relates the purge requirement to a single parameter, the Henry constant H. We shall show in subsequent sections how these expressions can be corrected in simple fashion to account for nonequilibrium effects.

2.2.2   Breakthrough Curves

Equations (2.3) and (2.4) can also be recast in a form suitable for calculating the so-called breakthrough curves, which log the time-variations of the fluid concentrations at the bed exit, i.e., Y_L = f(t). We write

$$\Delta q/\Delta Y = \frac{\rho_f v}{\rho_b \cdot L} t \quad \text{Adsorption} \tag{2.9}$$

$$dq/dY\big|_Y = \frac{\rho_f v}{\rho_t \cdot L} t \quad \text{Desorption} \tag{2.10}$$

In the case of adsorption, $\Delta q/\Delta Y$ is a constant, and all concentrations between the feed Y_F and the initial bed concentration Y_I break through simultaneously, i.e., the breakthrough curve is a discontinuous one. For desorption, each isotherm slope dq/dY is associated with a particular fluid concentration Y which can be read off the isotherm plot or calculated from the isotherm equation. In this case an elongated desorption breakthrough curve results, which can be calculated from Equation (2.10) and the given isotherm data. We demonstrate the necessary calculations in the following illustration.

Illustration 2.2: Calculation of a Desorption Breakthrough Curve Under Equilibrium Conditions

We consider a particular system, that of *m*-xylene in octane solvent in contact with a zeolite sorbent. Its equilibrium isotherm of type I (Langmuir) has been reported as q = 5.5 Y/(1 + 29Y), where Y is given in kilograms *m*-xylene per kilogram octane, and q in kilograms *m*-xylene per kilogram zeolite.

The derivative we require is then given by

$$dq/dY = 5.5/(1 + 29Y)^2$$

We set v = 10^{-3} m/s, a typical value for liquid systems, and bed height L at 0.4 m. The density ratio ρ_f/ρ_b is, for simplicity, set at unity, and the bed is assumed saturated at a fluid concentration of Y = 0.02.

We then obtain from Equation (2.4)

$$t = \frac{\rho_b}{\rho_f} \frac{L}{v} \frac{dq}{dY} = 1 \cdot \frac{0.4}{10^{-3}} \frac{dq}{dY} = 400 \frac{dq}{dY}$$

With these expressions in hand, the following numerical results are obtained

Y	0	0.001	0.005	0.01	0.015	0.020
dq/dY	5.5	5.19	4.20	3.13	2.67	2.20
t(s)	2200	2076	1680	1252	1068	881

Thus, the feed concentration of Y = 0.02 emerges from the desorbing bed after 881 s and desorption is complete after 2200 s, yielding a fairly elongated desorption breakthrough curve.

By contrast, adsorption breakthrough of the feed Y = 0.02 onto a clean bed occurs at

$$t = \frac{\rho_b}{\rho_f} \cdot \frac{L}{v} \cdot \left(\frac{d}{Y}\right)_F = 1 \cdot \frac{0.4}{10^{-3}} \cdot \frac{1}{1.58} = 253 \text{ s}$$

The desorption step is thus seen to be approximately ten times slower than the saturation step.

2.3   GENERAL COLUMN OPERATION

Graphical presentations of column operations are best made in terms of operating diagrams which, in subsequent figures, will be shown on the left and serve to convey a graphical picture of both the variables and the operation itself. On the right we shall display concentration profiles and histories (breakthrough curves) which derive from the operating diagrams.

The two sets of diagrams are followed by algebraic expressions which represent the mass balances, derived in the previous Section 2.2, and convey the following information:

> "Design equation", which (for lack of a better term) inter-relates the system parameters: time t, distance along the column z, mass velocity (G_b), or its equivalent, superficial velocity times fluid density $v\rho_f$, bed and fluid density (ρ_b, ρ_f), and expressions which describe the equilibrium relation. Propagation velocity V is given by the ratio z/t.

> "Breakthrough time" (adsorption) and "desorption time" (desorption), to convey a practical sense of the response of the system, it being understood that we are limiting our-selves to equilibrium conditions. The effect of mass transfer on the operation will be taken up in a subsequent section and our approach will be to graft its influence onto the results of our present equilibrium analysis.

> "Minimum bed requirements" (adsorption) and "mini-mum purge requirements" (desorption). These are the important design parameters which, above all others, determine the fixed and operating costs of the sorption operation. We speak of minimum quantities because of the constraints of our present equilibrium analysis. When mass transfer effects are taken into account, these minimum quantities given by Equations (2.6) and (2.8) will be enhanced by appropriate factors which will be taken up in a separate section. One should not, however, underestimate the value of the present results, which yield an extremely useful lower limit to the quantities in question, much as do the minimum solvent or reflux ratios in other mass transfer operations.

As we have seen the design equation can also be used to derive entire bed profiles and breakthrough curves under equilibrium conditions. For the former, one sets t = constant, for the latter z =

constant. Discontinuous fronts, such as arise in adsorption along a type I isotherm, yield a single value of z or t. Expanding fronts, on the other hand, form continuous profiles and breakthrough curves, which are characterized by the local slopes of the equilibrium curve. They are seen during desorption from a type I isotherm and adsorption along a type III equilibrium curve.

We now turn to a detailed consideration of column behavior for a number of different isotherm types.

2.4   COLUMN SORPTION FOR TYPE I ISOTHERMS

It is appropriate to start our considerations with this, the most frequently encountered isotherm form. Feed conditions to the adsorber bed are assumed constant and the adsorbent is taken to be uniformly preloaded (or clean) at the start of the operation. We address both adsorption and desorption operations and the analysis takes the form of simple graphical constructions and algebraic expressions (Figures 2.2 and 2.3).

Mass transfer effects are omitted for the time being, although we prepare the reader by sketching in actual behavior which arises from the presence of a mass transfer resistance. In general, however, the focus will be on what has become known as equilibrium theory, i.e., bed behavior is presented in terms of equilibrium concentration profiles or equilibrium concentration histories (breakthrough curves).

2.4.1   Adsorption

- The propagation velocity of the concentration front here is seen to be inversely proportional to the slope of the operating line. This means that a low slope $\Delta q/\Delta Y$, i.e., a high feed concentration, will result in fast movement of the front through the sorbent. This is in line with physical reasoning, i.e., a high feed concentration will lead to rapid saturation of the sorbent.

- High feed (fluid) flow rate also leads to fast bed saturation and low breakthrough times. This is again in accord with physical reasoning.

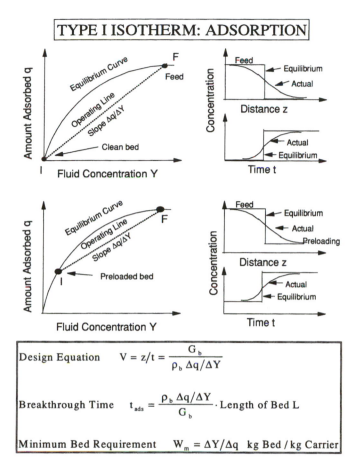

FIGURE 2.2
Operating diagrams (left) and concentration profiles/histories (right) for adsorption along a type I (Langmuir) isotherm. Relevant equations are summarized at bottom.

- The weight of sorbent required to treat a feed of concentration Y_F, q_F is given by the inverse of the operating line slope $\Delta q/\Delta Y$ in units of kilogram sorbent per kilogram carrier fluid passed through. Put in another way, the lower the slope of the operating line, the higher the required weight of sorbent. This important parameter can therefore be established in simple fashion from the operating diagram.

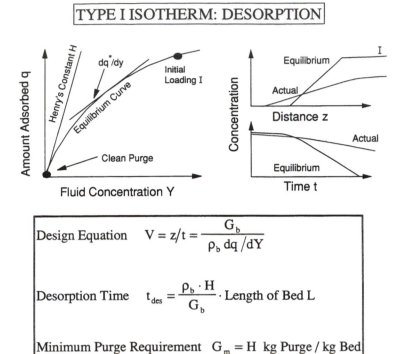

FIGURE 2.3
Desorption from a type I (Langmuir) isotherm.

2.4.2   Desorption

- Here the propagation velocity of a particular concentration level is inversely proportional to the slope of the equilibrium isotherm at that point. This implies, as in the case of adsorption, that the lower the slope dq/dY, the faster the movement of the concentration in question. This in turn leads to the result that the initial parts of a loaded sorbent which have a low slope desorb relatively quickly, while the final portions, particularly those near or at the origin, lead to long, trailing zones.

- As was pointed out previously, adsorption and desorption fronts differ drastically even under the present equilibrium conditions. The former lead to sharp discontinuities which depend only on the end-point concentrations of the operating line. In desorption, each point of the equilibrium curve exerts an effect: the profiles and histories become quasi-mirror images of the isotherm, steep portions of the concentration fronts corresponding to low slope segments of the isotherm, and vice versa. We thus obtain a continually broadening front, each point of which moves at a speed inversely proportional to the isotherm slope dq/dY. This is an undesirable condition, since the front will occupy a considerable portion of the bed and lead to a lengthy desorption process.

- The minimum weight of clean purge required, G_m, does NOT, remarkably, depend on the initial bed loading. It is, instead, uniquely determined by the isotherm slope at the end point of the desorption process, i.e., the Henry constant. We repeat the relevant equations for emphasis:

$$G_m = dq/dY \text{ for incomplete purge;}$$
$$G_m = H \text{ for complete purge}$$

Illustration 2.3: Minimum Bed Volume Requirement

One thousand cubic meters of air containing toluene at $Y = 0.01$ kg/kg air is to be treated in a carbon bed. The isotherm is of Langmuir form and has been reported as $q_{tol.} = \dfrac{2000\,Y}{1+2200\,Y}$.
Hence,

$$\frac{\Delta q}{\Delta Y} = \frac{q_F - q_I}{Y_F - Y_I} = \frac{q_F - 0}{Y_F - 0} = \frac{200 \cdot 0.01}{1 + 2200.01} = \frac{2}{23}$$

The minimum bed weight is (see Equation 2.6)

$$\frac{\text{Bed weight}}{\text{Weight of carrier}} = \frac{1}{\Delta q / \Delta Y}$$

or

$$\frac{\text{Bed volume} \cdot \rho_b}{1000 \cdot \rho_f} = \frac{23}{2}$$

Setting $\rho_f = 1\ kg/m^3$ and ρ_b for carbon $= 450\ kg/m^3$ (see Table 1.1) we obtain

$$\text{Minimum bed volume } = \frac{1}{450} \cdot 1000 \cdot \frac{23}{2} = 12.8\ m^3$$

Illustration 2.4: Matching Purge Time With Breakthrough Time

In Illustration 2.1, dealing with the system *m*-xylene/octane/zeolite, a breakthrough time of 253 s was obtained for $v = 10^{-3}$ m/s and L = 0.4 m. The purge step, in contrast, had a duration of 2200 s. We now wish to speed up the desorption process so that the two columns — one on stream, the other being regenerated — can be used in synchronized fashion. Henry's constant for the system was 5.5.

We recast the design equation in the form

$$v = \frac{\rho_b}{\rho_f} \frac{L}{t} H \text{ and obtain } 1 \cdot \frac{0.4}{253} \cdot 5.5 = 8.7 \cdot 10^{-3}\ m/s$$

A purge velocity 8.7 times higher than the feed velocity of $v = 10^{-3}$ m/s would thus have to be used to match the adsorption step.

2.5   COLUMN SORPTION FOR TYPE III ISOTHERMS

Since type III isotherms can be regarded as inversions of their favorable counterpart, type I, one would expect an exact reversal in sorption behavior. This is indeed the case: adsorption along a type III curve leads to an expanding front whose propagation speed varies inversely with the isotherm slope (Figure 2.4). It is low at the inlet, i.e., at feed conditions and progressively increases as one penetrates into the bed. This contrasts with the discontinuous adsorption front one obtains along type I isotherms.

For desorption, the roles are again reversed. It is now the type III isotherm which yields a sharp discontinuity, while type I leads to expanding fronts (Figure 2.5).

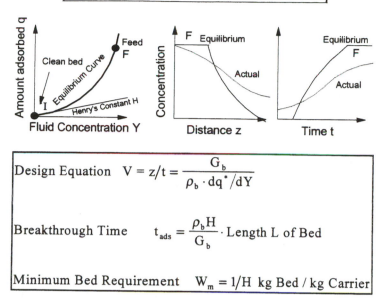

FIGURE 2.4

Adsorption along a type III (unfavorable) isotherm. Behavior is the reverse of that seen along a type I isotherm.

We shall see in a later section that type I isotherms, which perform poorly during desorption, can be converted into effective type III equilibrium curves by applying a hot gas purge for the regeneration. One thus has the double advantage of an efficient adsorption step and an equally efficient desorption step, both yielding sharp or discontinuous fronts.

2.6   COLUMN SORPTION FOR LINEAR ISOTHERMS

We are now in a position to make certain statements about systems with linear isotherms. We do this by considering the linear isotherm to be a limiting case of type I and type III isotherms, and by approaching that limit from either side by progressively reducing the curvature of those isotherms. The crucial parameters are the slope of the operating line $\Delta q/\Delta Y$ and the local slope of the isotherm dq/dY. For adsorption along a type I isotherm, $\Delta q/\Delta Y$ remains constant with diminishing curvature

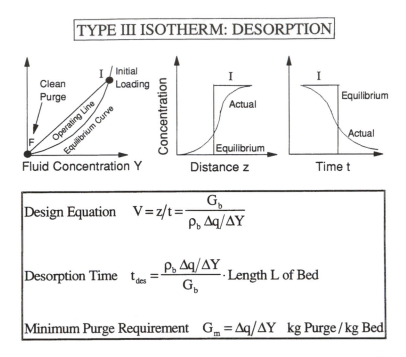

TYPE III ISOTHERM: DESORPTION

Design Equation $$V = z/t = \dfrac{G_b}{\rho_b \, \Delta q/\Delta Y}$$

Desorption Time $$t_{des} = \dfrac{\rho_b \, \Delta q/\Delta Y}{G_b} \cdot \text{Length L of Bed}$$

Minimum Purge Requirement $G_m = \Delta q/\Delta Y$ kg Purge / kg Bed

FIGURE 2.5
Desorption from a type III (unfavorable) isotherm. Types of profiles/histories are mirror images of those seen in adsorption along a type I isotherm.

of the isotherm, and upon reaching the linear limit, becomes identical to the Henry constant H. In a similar fashion, adsorption along a type III (unfavorable) isotherm ultimately leads, in the limit, to the condition $dq/dY = H$. For desorption the roles are reversed, dq/dY being associated with the type I isotherm, $\Delta q/\Delta Y$ with type III equilibrium curves. A similar limiting procedure leads to the same conditions: $\Delta q/\Delta Y = dq/dY = H$. We note that in approaching the limit $\Delta q/\Delta Y$ consistently maintains a discontinuous sorption front while dq/dY leads to profiles which become progressively sharper, converging to a discontinuity in the limit. We conclude from this that linear isotherms produce discontinuous fronts for both adsorption and desorption operations. This is in contrast to the results of type I and type III isotherms, which lead to both discontinuous fronts (adsorption along type I, desorption along type III) as well as expanding fronts (desorption along type I, adsorption along type III). The equations we had given for those cases retain their form, but with Henry's constant H taking the place of $\Delta q/\Delta Y$ and dq/dY. They are reproduced below for convenience:

Design equation

(adsorption and desorption)

$$V = z/t = \frac{G_b}{\rho_b \cdot H} \quad (2.11)$$

$$\left.\begin{array}{r}\text{Breakthrough time} \\ \text{desorption time}\end{array}\right\} \quad t = \frac{\rho_b \cdot H}{G_b} \times \text{length of bed L} \quad (2.12)$$

Minimum bed requirement $W_m = 1/H$ kg bed/kg carrier

$$(2.13)$$

Minimum purge requirement $G_m = H$ kg purge/kg bed

$$(2.14)$$

Illustration 2.5: Sorption of an Aqueous Pollutant from Groundwater

The situation considered involves the adsorption of a single solute from slow-moving groundwater (v = 1 mm/s) onto soil. The equilibrium loading for $Y = 10^{-4}$ kg/kg water has been determined at $q = 10^{-1}$ kg/kg soil, but it is not known whether the isotherm is favorable or linear. It is not expected that it will be unfavorable. Specific gravity of soil is set at the standard value of 2.5.

Operating line — The slope of this line spanning the isotherm between feed point and initial loading (= 0, i.e., initially clean medium) is given by

$$\frac{\Delta q}{\Delta Y} = \frac{10^{-1} - 0}{10^{-4} - 0} = 10^3 \frac{\text{kg water}}{\text{kg soil}}$$

This holds irrespective of whether the isotherm is linear or favorable. For linear equilibrium, $\Delta q/\Delta Y = H$ (Henry constant).

Propagation velocity of sorption front — This is given by the expression

$$V = \frac{z}{t} = \frac{G_b}{\rho_b \cdot \Delta q / \Delta Y} = \left(\frac{\rho_f}{\rho_b}\right) \cdot \frac{v}{\Delta q / \Delta Y}$$

Hence

$$V = 0.4 \cdot \frac{10^{-3}}{10^3} = 4 \cdot 10^{-7} \, \text{m/s}$$

Position after 100 days — From $V = z/t$ we obtain

$$\text{Position } z = t \cdot V = 100 \cdot 24 \cdot 3600 \cdot 4 \cdot 10^{-7}$$

$$z = 3.5 \, \text{m}$$

Soil in contact with pollutant — This is equivalent to "minimum bed requirement" and amounts to

$$W_m = 1/(\Delta q / \Delta Y) = 1/10^3 = 10^{-3} \, \frac{\text{kg soil}}{\text{kg water}}$$

Illustration 2.6: Clearance of Contaminated Soil

Desorption time — The soil contaminated in the previous example is cleared during the subsequent passage of clean groundwater. What is the duration of clearance assuming contamination lasted 100 days?

The calculation requires a Henry constant for both the linear and nonlinear isotherm. We choose $H = 10^4$ kg water/kg soil, typical of the higher polychlorinated biphenyls (Table 7.1). Note that the Henry constant must perforce be equal to or greater than the slope of the operating line. We obtain

$$t_{des} = \rho_b \cdot \frac{H}{G_b} L = \left(\frac{\rho_b}{\rho_f}\right) \cdot \frac{H}{v} L$$

$$t_{des} = 2.5 \cdot \frac{10^4}{10^{-3}} 3.5 = 8.8 \cdot 10^7 \, \text{s} = 1019 \, \text{days}$$

The fact that close to 3 years are required to clear a stretch of 3.4 m even under equilibrium conditions attests to the dangerous nature of these contaminants.

Desorption of a 10-fold contamination load — No change occurs in the clearance time, since the same Henry constant applies, irrespective of initial loading.

Illustration 2.7: Recovery of Solvent Vapour from Air

Ten parts per million by weight of a solvent vapor in air is to be recovered by adsorption on activated carbon. The operation is isothermal (see Chapter 4, on Adiabatic Sorption) and the isotherm can be assumed linear over the range in question. Henry constants for such vapors are typically in the range $H = 10$ to 10^3 kg air/kg carbon. We choose $H = 10$ and set fluid velocity at $v = 1$ m/s, bulk density at $\rho_b = 440$ kg/m^3, air ρ_f at 1 kg/m^3.

Bed length for a 10 h cycle time — The bed will be on stream for adsorption for 10 h, followed by regeneration with steam and cooling. We use the design equation

$$z = \frac{G_b t}{\rho_b \cdot H} = \left(\frac{\rho_f}{\rho_b}\right) \cdot \frac{vt}{H}$$

and obtain

$$z = (1/440) \cdot 1 \cdot (10 \cdot 3600)/10 = 8.2 \text{ m}$$

A bed of 8.2 m is required for a 10 h adsorption step or one of 82 cm depth for a 1 h step. In actual practice bed heights vary from 1 to 2 m, and on-stream adsorption times of 2 h are common.

2.7 COLUMN SORPTION FOR INFLECTING ISOTHERMS

Hitherto we had dealt with equilibrium curves that were free of inflections. These isotherms gave rise to concentration fronts which were either discontinuities or were uniformly broadening. One might surmise from this that, with the introduction of inflections, which contain both "favorable" and "unfavorable" segments, the fronts will consist of combinations of discontinuities and expanding fronts.

This is indeed the case, but it is not entirely clear where one segment ends and the next one begins. It is tempting to conclude that all segments that are favorable or unfavorable *at sight* should behave like type I or type III isotherms over their entire length. It will be quickly found, however, that this approach leads to discontinuities in the slope at the point of juncture, which in turn causes discontinuities in the velocities of the front, which is unacceptable on physical grounds. The proper approach is one that avoids this dilemma, and is best expressed in terms of what we call the IF rule, which is stated as follows:

Rule 1: IF one proceeds from the point of initial loading to the feed point, I → F, the slope of the trajectory must either increase along an equilibrium curve or, failing that, stay constant along an operating line.

It is quickly seen that the results we have presented so far agree with the rule: for adsorption on type I isotherms, for example, the trajectory from I to F had a constant slope equal that of the operating line (Figure 2.2). If one had chosen to follow the isotherm, a path of diminishing slope would have resulted, violating Rule 1. In desorption, on the other hand, we are able to follow the isotherm, since it increases in slope as we go from initial loading to feed (i.e., purge, see Figure 2.3).

2.7.1 Adsorption (Figure 2.6)

- The isotherm is composed of "favorable" and "unfavorable" portions. Note that the latter terminates *below* A, not at A. To span the trajectory I → F we apply the IF rule, which dictates that the slope must either increase or stay constant. This is accomplished, and can only be accomplished, by drawing a tangent from the origin (the initial loading point I) to the isotherm (point A). The trajectory is then divided into a discontinuous front from I to A, followed by a broadening front from A to the feed point F.
- Both the general design equation and the breakthrough time retain their previous forms, but are allowed to use

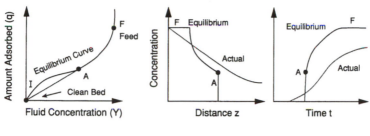

TYPE II ISOTHERM: ADSORPTION

$$V = z/t = \frac{G_b}{\rho_b(\Delta q/\Delta Y)_A} \quad OR \quad \frac{G_b}{\rho_b(dq\ /dY)_A}$$

Breakthrough Time $\quad t_{ads} = \dfrac{\rho_b(\Delta q/\Delta Y)_A}{G_b} \cdot$ Length L of Bed

Minimum Bed Requirement $\quad W_m = (\Delta Y/\Delta q)_A \quad$ kg Bed / kg Carrier

FIGURE 2.6
Adsorption along an inflecting type II isotherm. Both broadening and discontinuous fronts make their appearance. Relevant equations are summarized at the bottom of the figure.

> either of the slopes of the discontinuity $\Delta q/\Delta Y$ or that of the broadening front dq/dY, which are identical at the point of contact.

- The minimum bed requirement is dictated by the discontinuous portion of the front which is the first to emerge on breakthrough. Later portions of the trajectory have no effect on either breakthrough time or bed requirements.

- If the feed point F is moved upward, the general type of the front will be retained. If it is moved downward, the broadening portion will shrink and ultimately disappear when F = A. Further lowering of F will result in a single discontinuous front, as would result from a type I isotherm. It shrinks in length as F is reduced.

2.7.2   Desorption (Figure 2.7)

- Most aspects here correspond to the adsorption case. Again the trajectory is prescribed by the IF rule and results in a tangent drawn from the initial loading point I to the isotherm, leading to a discontinuous front IA and a broadening one from A to the origin.

- Operational requirements (purge consumption) depend solely on the Henry constant and are not affected by other portions of the isotherm. In particular, they are independent of the initial loading of the bed.

- A movement of I toward the origin will reduce the length of the discontinuous front which ultimately disappears completely upon reaching A. We are then left with a single broadening front, much as in the case of desorption from a type I isotherm.

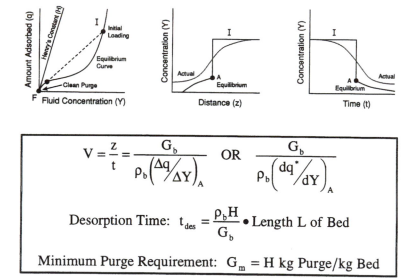

FIGURE 2.7

Desorption along an inflecting type II isotherm.

Illustration 2.8: Choice of Bed Diameter

We consider a flow rate of $Q = 10$ m³/s, roughly equivalent to 20,000 SCFM (standard cubic feet per minute), and choose a design velocity $v = 1$ m/s. We obtain for the bed area

$$A = \frac{Q}{v} = \frac{10}{1} = 10 \text{ m}^2$$

The following column diameters would provide the required 10 m² of cross-sectional area.

Number of beds	1	3	5
Diameter of bed (m)	3.6	2.1	1.6

All of these can be accommodated within industrial practice. The actual choice will depend on column cost and operating pressure. Thus, for high pressure operation, a set of smaller diameter columns would be preferred. Note that both Q and v diminish in the same proportion with elevation of pressure. The calculations given here for standard atmospheric conditions therefore remain valid.

Illustration 2.9: Drying of a Carbon Bed

Carbon has a type III water isotherm up to relative humidities of 50% to 75% (see Figure 1.3). Here adsorption fronts are of the broadening type, while desorption, which is the process considered, produces a sharp front.

Suppose a carbon bed of $z = 1$ m height has been saturated at 50% air humidity. What would be the duration of a drying process based on throughflow of air at $v = 1$ m/s?

Desorption time — At this stage we are only able to compute the *maximum* time under isothermal conditions. Fifty percent relative humidity translates into 0.01 kg H_2O/kg air, and the loading at this value is $q = 0.1$ kg H_2O/kg carbon. Hence, using a bone dry purge we have an operating line slope of $\Delta q/\Delta Y = 0.1/0.01 = 10$. Desorption time is then given by

$$t_{des} = \left(\frac{\rho_b}{\rho_f}\right)\frac{\Delta q/\Delta Y}{v} \cdot L = \frac{440}{1} \cdot \frac{10}{1} \cdot 1 = 4.4 \cdot 10^3 \text{ s}$$

Air consumption — This is immediately given by

$$W_m = 1/\Delta q/\Delta Y) = 0.1 \text{ kg air/kg bed}$$

Note that both desorption time and air consumption will be reduced at higher air temperatures. Desorption times of a few minutes would then not be uncommon.

Illustration 2.10: Adsorption Along an Inflecting Isotherm

Inflecting isotherms arise with some frequency in the adsorption of vapors on various solids, particularly at low temperature. A less well-known occurrence is in the field of adiabatic sorption. Both adsorption and to a lesser degree thermal desorption, often proceed along inflecting pathways which we have termed "effective equilibrium curves". Examples of such curves appear in Figures 4.5 and 4.6, and methods for deriving them are presented in Chapter 4.

Our purpose here is to establish pertinent adsorption parameters along one such inflecting isotherm. We choose the pathway describing adsorption at 100% relative humidity and 100 kPa onto silica gel at 25°C (Figure 4.5). A tangent is drawn from the feed point to the isotherm. This creates a discontinuous front with $\Delta q/\Delta Y = 50$ kg air/kg solid. Beyond the tangent point, the path continues down a broadening front along the isotherm and terminates at a Henry constant of $H \cong 2$ kg air/kg solid (Figure 4.5). The sorption front consequently consists of a broadening front which breaks through first, followed by a discontinuous front which brings up the trailing end. We choose a bed height of $z = 2$ m and set velocity at $v = 1$ m/s, ratio of densities $\rho_b/\rho_f = 10^3$.

Breakthrough time — This follows from the cited equations. We have

$$t_{ads} = \frac{\rho_b H}{G_b} = \left(\frac{\rho_b}{\rho_f}\right) \cdot \frac{H}{v} L = 10^3 \cdot \frac{2}{1} \cdot 2 = 4000 \text{ s}$$

Bed requirement — The minimum bed requirement here is the inverse of the Henry constant, yielding

$$W_m = 1/H = 1/2 = 0.5 \text{ kg bed/kg carrier}$$

Length of transfer zone — We consider conditions after 1 h of operation and compute the distance, measured from the inlet, of the leading broadening front and the trailing end of the discontinuous section:

$$\textit{Trailing end}: \ z = \frac{G_b t}{\rho_b \, \Delta q/\Delta Y} = \left(\frac{\rho_f}{\rho_b}\right)\frac{vt}{\Delta q/\Delta Y} = 10^{-3} \cdot \frac{1 \cdot 3600}{50} = 0.072 \text{ m}$$

$$\textit{Leading front}: \ z = \frac{G_b t}{\rho_b H} = \frac{\rho_f}{\rho_b}\frac{vt}{H} = \frac{10^{-3} \cdot 1 \cdot 3600}{2} = 1.8 \text{ m}$$

The transfer zone is seen to occupy almost the entire bed: Δz = 1.8 – 0.07 = 1.73 m, compared to a total bed length of 2 m. It would be even longer under nonequilibrium conditions. Bed requirement is also inordinately high at 0.5 kg bed/kg carrier. The process is thus highly inefficient and alternatives must be sought.

Remedies — One preferred version is to raise the system (and inlet feed) pressure. This has two desirable consequences: temperature effects are attenuated, and the effective equilibrium curves shift toward the ordinate, yielding more favorable Henry constants and operating lines. We consider the case depicted in the diagram, Figure 4.5, adsorption of saturated water vapor from air onto silica gel at 500 kPa (~5 atm) and 25°C. There is an evident shift of the adiabatic equilibrium curves to the left. The Henry constant is now estimated at H = 70 kg air/kg bed, and the slope of the operating line of the discontinuous front at $\Delta q/\Delta Y$ = 100 kg air/kg bed. We obtain a minimum bed requirement and the length of the transfer zone.

Minimum bed requirement

$$W_m = 1/H = 1/70 = 0.014 \text{ kg bed/kg air}$$

Length of transfer zone

$$\text{Trailing end}: \ z = \left(\frac{\rho_f}{\rho_b}\right) \cdot \frac{vt}{\Delta q / \Delta Y} = \frac{10^{-3} \cdot 1 \cdot 3600}{100} = 0.036 \text{ m}$$

$$\text{Leading front}: \ z = \left(\frac{\rho_f}{\rho_b}\right) \cdot \frac{v \cdot t}{H} = \frac{10^{-3} \cdot 1 \cdot 3600}{70} = 0.051 \text{ m}$$

Thus, the zone length has shrunk from 1.8 m to less than 2 cm, and bed requirements have been reduced by almost two orders of magnitude, by merely raising the pressure from 1 to 5 atm. This surprising result may be explained as follows: an increase in total pressure causes a proportional shift of the *isotherms* toward the ordinate, since Y varies inversely with total pressure at constant humidity. Thus, the same loading results in one fifth the value of Y. The adiabatic equilibrium curves experience a similar shift, but the effect cuts more deeply: there is a simultaneous change toward a more "favorable" shape of equilibrium curve, which causes a disproportionate increase in Henry's constant from 2 to 70. It is this increase which causes the dramatic drop in transfer zone length.

Chapter 3

EQUILIBRIUM EFFECTS: THE BINARY LANGMUIR CASE

With the introduction of a second component an immediate escalation in complexity occurs. One is now dealing with two equilibrium diagrams, each one of which contains not a single isotherm but a network of what we call partial isotherms or "characteristics" which we denote q_i^+ and q_i^-.

This construction is necessary since one is now forced to accommodate binary compositions, denoted (Y_1, Y_2) in the diagram. We chose the subscript 1 to denote the light component (lowest Henry constant), subscript 2, the "heavy" one.

Some equations are necessary to explain the genesis of these diagrams. The equilibrium relation itself is represented by the binary Langmuir isotherm, which takes the form

$$q_1 = \frac{H_1 Y_1}{1 + b_1 Y_1 + b_2 Y_2} \tag{3.1}$$

$$q_2 = \frac{H_2 Y_2}{1 + b_1 Y_1 + b_2 Y_2} \tag{3.2}$$

These isotherms, although of elementary form, yield a surprisingly good picture of the general behavior of binary systems and are therefore retained for general instructional purposes.

For the generation of the characteristics we must, with apologies, resort to a differential equation, which is reproduced below:

$$\frac{dY_2}{dY_1} = \frac{q_{22} - q_{11} \pm \left\{ \left[q_{22} - q_{11} \right]^2 + 4 q_{12} \cdot q_{21} \right\}^{1/2}}{2 q_{12}} \tag{3.3}$$

Here q_{11}, q_{22}, q_{12} are the partial derivatives of the Langmuir isotherms, the first index giving the component number, and the second index the direction of differentiation. Thus, for example, $q_{12} = \partial q_1 / \partial Y_2$.

We note that the differential equation is a quadratic and gives rise to two sets of solutions $Y_2 = f(Y_1)_+$ and $Y_2 = f(Y_1)_-$. Upon substitution into the Langmuir isotherm these functions yield two sets of characteristics which we need for our operational diagrams $q_1 = f(Y_1)$ and $q_2 = g(Y_2)$. For our purposes, we merely note the shape of these curves: q_1^+, q_1^-, and q_2^- all emanate from the origin, q_1^- and q_2^- intersecting the pure component isotherm. q_2^+ does not pass through the origin but intersects the pure component isotherm (Figure 3.1). We are content with this qualitative picture, since we shall merely use the diagrams to deduce some general properties of binary systems, irrespective of the precise location of a particular characteristic, feed point F or initial loading point I.

The second feature of note is the existence of a point W, the so-called watershed point, which corresponds to the double root of the differential equation, i.e., ± give the same value. This point influences the evolution of binary fronts and in effect divides the operational results into two domains of drastically different behavior (hence the term watershed point). We shall enter into this later.

We note here that we retain the characteristics only very briefly to convey a picture of possible sorption pathways and enshrine the results in certain rules. Thereafter we shall relinquish the characteristics and merely make use of the rules and the pure component isotherms. In other words, we shall have no need of the binary Langmuir isotherms, Equations (3.1) and (3.2), or for that matter of the differential Equation (3.3). All we shall require to trace the complete sorption trajectories in many important cases will be the location of feed and initial loading points, and the use of the pure component isotherms. This represents an immense simplification of the analysis of binary sorption operations. In many instances we shall be able to extend our analysis to isotherm forms other than type I and in the process reveal hitherto unknown or unsuspected types of behavior which would go unnoticed in the usual computer simulations (see in particular Section 3.3, Mutual Displacement of Components).

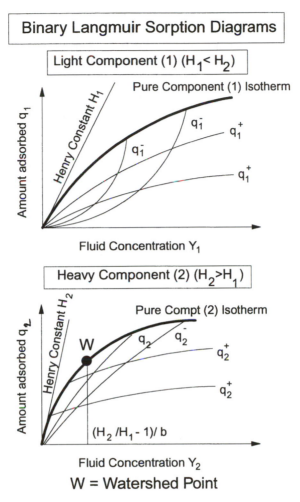

FIGURE 3.1

Characteristics (pathways) for binary Langmuir sorption. Feed F is located on negative characteristics q_1^-, q_2^- and the initial bed condition I on positive characteristics q_1^+, q_2^+. The watershed point W divides the graphs into domains of different operational behavior.

3.1 BINARY ADSORPTION AND DESORPTION

We now present a number of graphical constructions (the "doodling" referred to in the introduction), without concerning ourselves with the precise location of the parameters or the trajectories. At this point we still retain the characteristics.

In this first presentation we address two classical cases, the adsorption of a binary feed and the desorption of a binary mixture from a bed. In this we make use of a new rule, Rule 2 (in addition to the IF Rule), which is formulated below:

Rule 2:   One departs from the initial loading point I on a q^+ characteristic, and arrives at the feed point F on a q^- characteristic.

Keep in mind that in tracing out the trajectory between I and F, the slope has to increase along an equilibrium curve or, failing that, stay constant along an operating line. We find that in doing this for a binary system, an intersection will occur between the two paths connecting I and F. The slopes at the juncture will differ, with the result that one end-point concentration will lag, the other one will advance, leading to the opening up of a plateau of constant concentration which increases in length as the front penetrates into the sorbent matrix. Two such plateaus make their appearance in the two classical cases in the accompanying diagram (Figure 3.2).
Comments:

- Rule 2 is applied to establish the location of the starting and end points of the trajectory: F on a q^- and I on a q^+.
- To connect the two, Rule 1 (IF rule) is applied to trace out the complete trajectory. In the case of adsorption this leads to two discontinuous fronts. For desorption the path leads along characteristics and isotherms and results in two broadening fronts because of the increase in slope.
- The intersections of the two paths emanating from I and F lead to ever-expanding plateaus of a concentration corresponding to that of P.
- Desorption with pure purge of a column uniformly loaded with binary feed always leads to two expanding concentration fronts, separated by a plateau of pure heavy component 2 which has a concentration lower than that of W. The total desorption time is independent of the initial loading of the bed and is equal to

$$t_{des} = \frac{H_2 \rho_b}{G_b} \text{ length L of bed}$$

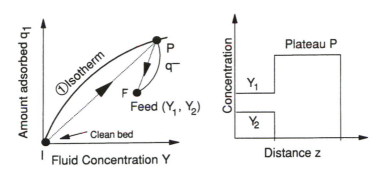

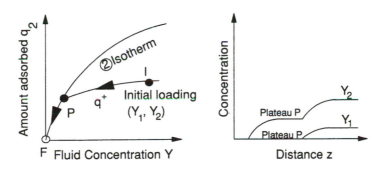

FIGURE 3.2

Operating diagrams and profiles for binary Langmuir adsorption and desorption. Pure component plateaus make their appearance in both cases.

It depends only on the Henry constant of the heavy component H_2, a feature we shall encounter again in multicomponent systems.

3.2 CREATIVE DOODLING: PREDICTION OF VARIOUS BINARY ADSORPTION PROFILES

Binary feeds can contact a bed under a variety of different conditions: the bed may be clean, or it may be preloaded with either heavy (2) or light (1) component, or even both. The principles we have established previously allow us to make predictions for these cases in simple graphical fashion on our usual q-Y operating diagrams without recourse to the detailed characteristics, which are now set aside. The only tools needed are the pure component isotherms and the location of the feed point.

The cardinal principles in this procedure are the two rules promulgated previously (Rules 1 (IF) and 2) and that mixed bands containing both components must proceed at the same propagation velocity, i.e., $\Delta q/\Delta Y$ (or dq/dY) must be identical for both components when they appear in unison. We demonstrate the principles involved for three cases (Figure 3.3):

(I)   A clean bed taking a binary feed
(II)   A bed preloaded with light component taking a binary feed
(III)   A bed preloaded with heavy component taking a binary feed

Comments:

- All three cases start by connecting the feed component that is not present in the preloaded bed with the origin. This has to be a straight operating line according to the IF Rule, since following the characteristic would yield a path of diminishing slopes. The second component which emerges at the same time has to have the same propagation velocity, hence the same slope or chord. These two initial moves in our doodling procedure yield two parallel lines through the feed points and establish all succeeding events: intersections are obtained with the pure component isotherms that define the plateaus. In most instances, this results in discontinuous jump changes in concentration, except when the trajectory follows an isotherm of increasing slope (case (II)) in which case a broadening front results in accordance with the IF Rule.

- A specific example will serve as an illustration. In case (II), the bed is preloaded with component 1, hence the "heavy" feed F(2) must contact the bed devoid of component 2 at the origin. This is done by a straight line chord from F(2) to the origin, to avoid violation of the IF Rule. A parallel line is next drawn through the point F(1). This establishes the plateau P on the (1) isotherm and at the same time

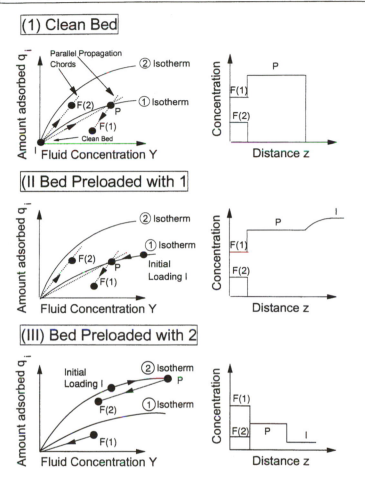

FIGURE 3.3

Graphical constructions for various types of binary Langmuir sorption. Operating lines through the feed points F(1), F(2) are parallel. Only pure component isotherms are required for the construction.

delineates the trajectory which descends from the initial loading point I to the plateau P, leading to an expanding concentration front. After passage of the plateau the path continues along a discontinuous front to its final destination, the feed point F(1). Simultaneously component 2 rises on a parallel line from its initial zero concentration to reach its feed value F(2).

3.3   MUTUAL DISPLACEMENT OF INDIVIDUAL COMPONENTS

3.3.1   The Watershed Point W

By mutual displacement of individual components we wish to convey situations in which a bed fully saturated with light component 1 receives a feed of heavy component 2 or, conversely, component 1 is used to displace the heavier component 2 from a uniformly and fully saturated bed. In these operations the so-called watershed point W turns out to be of crucial importance. We had briefly alluded to some of its properties, but it is now appropriate to present the full panoply of its characteristics, which are displayed in Figure 3.4.

- Mathematically, the watershed point W and its concentrations (q_{2w}, Y_{2w}) result from the special case when the quadratic differential Equation (3.3) exhibits a double root, much as simple second-order differential equations exhibit double roots which signal the transition from exponential to periodic solutions. In the case of the present differential equations, W signals the transition of the original model equations from so-called hyperbolic behavior to parabolic behavior. The exact meaning of these terms need not concern us, except that the change in mathematical form brings with it a change in the solution curves (much as they do in the case of second-order ordinary differential equations [ODEs]). This will be the main focus of our concern.

- The watershed point is located on the pure component 2 isotherm with an abscissa value of $Y_{w2} = (H_2/H_1 - 1)/b_2$ where $H_{1,2}$ are the Henry's constants of the pure component isotherms, and b is a parameter of the binary Langmuir Equations (3.1) and (3.2). The equation can be circumvented by the following equivalent and simple graphical construction: the pure component isotherms are plotted in the same q-Y diagram and a tangent drawn to the light component isotherm (1) at its origin (slope = H_1). Its intersection with the heavy (2) isotherm fixes the location of W with coordinates Y_{2w} and q_{2w}. The construction is not only simple but also capable of conveying an immediate sense of the location of W, and its consequences (see Figure 3.4).

- W has the additional property of dividing the pure component 2 isotherm into two segments: above W, the isotherm has

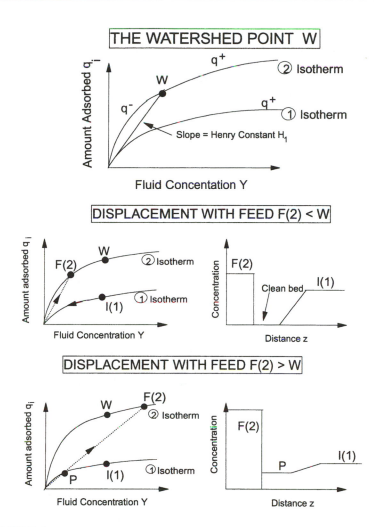

FIGURE 3.4
The watershed point W and its effect on the displacement of a light component (1) by a heavy component (2).

a q^+ label, below it, it becomes q^-. Since feed F and initial loading I have to be located on q^- and q^+, respectively, it is clear that this division will have an effect on the concentration fronts produced with different locations of F and I. These properties are again illustrated in Figure 3.4.

In tracing the concentration trajectories from the operating diagrams, several points have to be kept in mind:

- F has to be on a q^- characteristic, I on a q^+ one (Rule 2).
- The *shape* of the concentration fronts is determined by the IF Rule (Rule 1): trajectory slopes have to either increase (in which case a broadening front results) or stay constant (leading to discontinuous fronts).
- Finally, it must be kept in mind that the displacing agent (F) has to connect to its initial loading point which lies at the origin (Figure 3.4).

3.3.2   Displacement of (1) with F (2) < W

Both F and I are located on properly labeled isotherms (q^- for F, q^+ for I) so it becomes a matter of connecting the two points in appropriate fashion. This can only be done by passing through the origin of the q-Y plane; hence, the proper procedure is to descend from I to the origin (giving rise to a broadening front because of the IF rule) and proceed from there to the feed point F, which gives rise to a discontinuous front (again because of the IF rule). Because of the difference in propagation velocity of the two trajectories at the origin, a segment of "clean bed" opens up between the two which increases in length as the displacement process progresses. This is shown in the accompanying figure (3.4).

There is an important consequence in this development which is most clearly seen from the expression for the desorption time of component (1):

$$t_{des} = \frac{H_1 \rho_b}{G_b} \times \text{length of bed L}$$

We have seen these expressions before, but there is an additional implication here: t_{des} is independent of the concentration of the displacing agent F(2), i.e., no matter how high we push F (while staying below the watershed point), nothing will happen to speed up the displacement process. This flies in the face of intuition, but is nevertheless a well-established fact: below W the concentration of the displacing agent has no effect on the displacement process and never exceeds the results one would obtain by displacing with pure carrier fluid.

3.3.3 Displacement of (1) with F(2) > W

Things begin to change once we raise F(2) above the watershed point W. F(2) is now located on an isotherm segment labeled q^+ but must in fact be reached on a q^-. This has the effect of changing the trajectory leading to the feed point F(2) which no longer proceeds along the pure component 2 isotherms. Instead the sorption path evolves as follows.

As indicated in Figure 3.4, we descend from the initial loading point I and intersect with the trajectory through F(2) and the origin which according to the IF Rule has to be a straight line. The point of intersection occurs at the plateau P, with the consequence that we no longer have a segment of "clean bed" opening up between the two trajectories. This has the effect of speeding up displacement and increasing enrichment in the component (1): the higher F(2), the higher the plateau and the shorter the segment IP, making for faster desorption times.

Illustration 3.1: Adsorption of a Binary Mixture

The procedure laid down in the text showed three requirements for the evaluation of this case: equilibrium feed compositions for light (1) and heavy component (2), and the pure component isotherm for the light species. We consider the actual case of the adsorption of a binary mixture, m-xylene and p-xylene, from octane solvent onto a zeolite.

	F(1)	**F(2)**
Feed concentrations	$q_{F1} = 0.02$	$q_{F2} = 0.145$
	$Y_{F1} = 0.06$	$Y_{F2} = 0.06$
Langmuir equation for light component	$q_1 = \dfrac{5.5 \cdot Y_1}{1 + 29 \cdot Y_1}$	

Plateau of light component — Graphically, the plateau is established by drawing a line through the origin and F(2), which has the slope $q_{F2}/Y_{F2} = 0.145/0.06 = 2.4$. A line of the same slope is drawn through F(1), intersecting the light component isotherm at the plateau value $q_p = 0.144$ and $Y_p = 0.11$.

Analytically, the same result can be obtained by solving for the intersection of the light component isotherm with the operating line:

$$2.4 = \frac{q_{F1} - q_p}{Y_{F1} - Y_p} = \frac{0.02 - \dfrac{5.5\, Y_p}{1 + 29\, Y_p}}{Y_{F1} - Y_p}$$

This yields the same plateau value of $Y_p = 0.11$, and by substitution into the Langmuir isotherm equation, $q_p = 0.144$.

Enrichment factor — Plateau and feed concentrations allow us to calculate the enrichment, or elevation of the plateau concentration over that in the feed

$$E = \frac{Y_p}{Y_F} = \frac{0.11}{0.06} = 1.8$$

In actual practice an enrichment of approximately 1.5 is obtained because the mass transfer resistance erodes the plateau to a peak of lower concentration. We shall shortly address the methods available for minimizing transport effects and taking advantage of the full plateau values.

Recovery R of light component — By the time the second component breaks through, the bed will have been saturated with both light and heavy components, i.e., it is in equilibrium with the feed solution. One can use this point to compute the fraction of light solute in the plateau which has been recovered in pure dissolved form as a fraction of the total passed through. It is given by the ratio of plateau to feed concentrations, multiplied by a ratio of the appropriate operating lines connecting the origin and feed point to the plateau:

$$R = \frac{Y_p}{Y_F}\left[\frac{\left(\dfrac{\Delta q}{\Delta Y}\right)_{FP} - \left(\dfrac{\Delta q}{\Delta Y}\right)_{OP}}{\left(\dfrac{\Delta q}{\Delta Y}\right)_{FP}}\right] \tag{3.4}$$

where $\left(\dfrac{\Delta q}{\Delta Y}\right)_{FP} = 2.4$ and $\left(\dfrac{\Delta q}{\Delta Y}\right)_{OP} = \dfrac{0.144}{0.11} = 1.3$.

We obtain

$$R = \frac{0.11}{0.06}\frac{2.4-1.3}{2.4} = 0.84$$

Thus, 84% of the total feed in *m*-xylene is recovered in "pure" form, i.e., devoid of *p*-xylene. In practice, this value is again considerably lowered due to mass transfer resistance.

Breakthrough time — We use the experimental values for velocity $v = 2 \cdot 10^{-4}$ m/s and bed height $z = 0.4$ m and set $\rho_b/\rho_f = 1$ for simplicity. We obtain

$$t_{ads\,1} = \frac{\rho_b}{\rho_f} \cdot \frac{(\Delta q/\Delta Y)_1}{v} \cdot L = 1 \cdot \frac{1.3}{2\cdot 10^{-4}} \cdot 0.4 = 2.6\cdot 10^3\,s = 43.3 \text{ min}$$

$$t_{ads\,2} = \frac{\rho_b}{\rho_f} \cdot \frac{(\Delta q/\Delta Y)_2}{v} \cdot L = 1 \cdot \frac{2.4}{2\cdot 10^{-4}} \cdot 0.4 = 4.8\cdot 10^3\,s = 80.0 \text{ min}$$

Thus, under equilibrium conditions the light and heavy components break through after 43 and 80 minutes, respectively.

Illustration 3.2: Enrichment by Displacement

The graphical constructions given in the text so far serves us as a guide here: we had seen that the enrichment level is determined by drawing a straight line from origin to feed F (the displacement agent 2). Intersection with the component 1 isotherm determines the location of the plateau P, i.e., the concentration level at which the displaced component 1 leaves the bed (see Figures 3.4 and 3.5).

Of course, enrichment results only if F is raised to sufficiently high levels. If F is only modestly above the watershed point W, the initial loading point I will still drop, albeit not to zero. It is only when the line from feed to origin lies to the right of I that an enrichment in the loaded component is attained. The advantage of graphical constructions is that the necessary conditions can be instantaneously established.

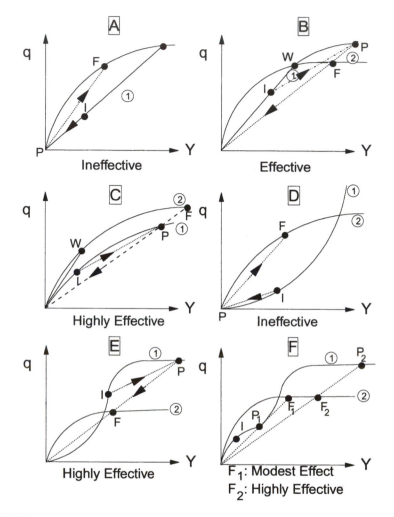

FIGURE 3.5
The effect of feed location and isotherm shape on the enrichment of a light component (1) to plateau values P when displaced by a heavy component (2). Only pure component isotherms are required in the construction.

We match the graphical construction by an analytical approach. The pertinent equations are as follows:

Straight line through plateau and feed $\quad \dfrac{q_2}{q_1 p} = \dfrac{Y_2}{Y_1 p}$ $\quad$ (3.5)

Langmuir equations for light component 1 and displacement agent 2

$$\frac{q_1}{Y_1} = \frac{H_1}{1 + b_1 Y_1} \tag{3.6}$$

$$\frac{q_2}{Y_2} = \frac{H_2}{1 + b_2 Y_2} \tag{3.7}$$

These can be combined into the single expression

$$Y_{2F} = \frac{H_2}{H_1} \frac{1 + b_1 Y_{1P}}{b_2} \tag{3.8}$$

where Y_{2F} = concentration of the displacement agent and Y_{1P} = enriched plateau concentration of the light component.

This expression reveals that in order to keep the consumption of displacement agent, i.e., Y_{2F} low, its Henry constant should approach that of the displaced species, i.e., $H_2 \approx H_1$. Thus, contrary to conventional wisdom, the most efficient displacement operations involve similar substances.

We consider the following numerical example:

$$q_1 = \frac{10 \cdot Y_1}{1 + 10\ Y_1}$$

Langmuir Isotherms

$$q_2 = \frac{20 \cdot Y_2}{1 + 10\ Y_2}$$

It is desired to enrich the light component by a factor of 10 from $Y_1 = 10^{-2}$ to 10^{-1} kg solute/kg carrier using a displacement technique.

Watershed point W — This is given by the analytical expression

$$Y_{2W} = \left(\frac{H_2}{H_1} - 1\right)/b_2 = (2-1)/10 = 0.1 \text{ kg solute/kg carrier}$$

Thus, below a concentration of $Y_{2W} = 0.1$, the displacement agent is ineffective, or no more effective than the carrier fluid.

Enrichment process — Here, the pertinent equation is that previously cited above (Equation 3.8). We have

$$Y_2 = \frac{H_2}{H_1}\frac{1+b_1 Y_{1p}}{b_2} = \frac{20}{10} \cdot \frac{1+10 \cdot 10^{-1}}{10} = 0.4 \text{ kg solute/kg carrier}$$

Thus, to obtain the desired tenfold enrichment, the displacement agent concentration must be 4 times the watershed concentration $Y_{2W} = 0.1$, or 40 times the original concentration of the displaced species, $Y_1 = 10^{-2}$.

The graphical procedures outlined in the text yield the same result but require a plot of the pure component isotherms.

Illustration 3.3: More Doodling — Displacement from Various Isotherms (Figure 3.5)

We consider here the displacement by a "heavy" component 2 with type I isotherm behavior of a light component 1 located on equilibrium curves of various shapes. Henry's constant $H_2 > H_1$ in each case. The guiding principle established for Langmuir solutes is the IF Rule, which requires that a straight line be drawn from feed to origin. Its intersection with the (1) isotherm locates the plateau.

Case A — Here the displacement agent 2 interacts with a light component 1 with a linear isotherm. The watershed point W is located at the intersection of the two equilibrium curves, and any feed concentration below that value fails to actively displace the light component. An ever-widening plateau P of clean bed opens up between two discontinuous fronts through I and F.

Case B — We recall that all linear isotherms must ultimately become one of the five types discussed, I–V. Here the linear isotherm of case A is extended to form a type I isotherm. The watershed point W is again located at the intersection of Henry's line with the displacing isotherm, and no active displacement of the light component 1 takes place below W. Once the concentration of the displacement agent rises above Y_{2W}, however, this situation is reversed and an enriched output of the light component at the plateau value P may be obtained.

Case C — As noted in the text, displacement can be rendered more effective and efficient by choosing a heavy component similar in isotherm shape to the light solute to be displaced. The reason is that the watershed point W is limited to low values, and a wide range of concentrations above Y_{2W} is therefore available to carry out the operation. High plateau values P may be attained, yielding substantial enrichment levels in the light component.

Case D — This unusual case considers the interaction of type I and type III isotherms. The watershed point W no longer plays its customary role, since the type III isotherm does not follow Langmuirian behavior. The IF Rule, however, still applies, and for the case shown indicates that no active displacement can take place. Two sharp fronts through I and F result, which flank an expanding section of clean bed, seen before under more conventional circumstances.

Case E — The type III isotherm of case D has been extended here, effectively becoming a type V isotherm reaching into a region of potential condensation or precipitation. Drawing a straight line through displacement feed F and the origin in accordance with the IF Rule, intersection with a plateau P of the light component 1 is obtained which is considerably enriched over the initial concentration I. One can visualize increased levels of Y_{2F}, ultimately leading to condensation or precipitation.

Case F — Here we consider interaction of the type I displacement agent with a light component of type IV. The behavior shows similarities to cases D and E. Below the watershed point W (not shown) no active displacement takes place.

Above it, only modestly high plateau levels are obtained until one reaches the threshold concentration F_1 lying on the tangent through the origin to the light component isotherm. F_1 still yields only modest enrichments here (see plateau P_1) but as soon as it is even slightly exceeded, a jump increase to much higher plateau levels occurs (see F_2 and P_2). It is evident that we are dealing with an exceptional phenomenon, akin to the "catastrophes" which occur in other physical systems on effecting a slight change in one of the system parameters. Further increases in F are again expected to result in phase separation in the light component as in case E, which can be exploited for easy recovery of that solute.

We have presented these examples largely in order to press the case for imaginative graphical constructions which are unsurpassed in their power to reveal new, unexpected features while retaining the facility to handle conventional operations. We must caution here that the straight line descent from P via F to the origin hinges on the q⁻ characteristic being concave upward. This was shown to be the case for a number of pure component isotherms of mixed type, but could not be proven in global fashion. Deviations would alter the rear zone to a partially or completely expanding one but in no way alters the fact that a substantial jump enrichment from I to some higher plateau P is attained.

Illustration 3.4: More on Displacement of a Type III by Type I Isotherm (Figure 3.6)

In case D of the preceding illustration we had briefly alluded to this case. Both feed point F and initial loading point I were set at relatively low levels and the path took the form of two discontinuous fronts (dotted lines), one descending from I to the origin, the other rising from the origin to F. Since they propagate at different speeds, a section of clean bed represented by the origin opens up between the two fronts.

We now proceed to move F, and ultimately I, to higher levels. In the first case shown in Figure 3.6, F is placed at the intersection of the two isotherms while I is kept at a low level. The IF Rule is obeyed in this case (i.e., the slope OF > slope OI) and the construction reverts to that of case D of the preceding illustration. We have sketched the corresponding profile, which shows the "clean bed" opening up between the two discontinuous fronts.

In the second case, F is moved beyond the isotherm intersection for both low and high levels of I. A reversal of slopes takes place, violating the IF rule. Clearly a descent to the origin can no longer take place, and alternative pathways must be sought.

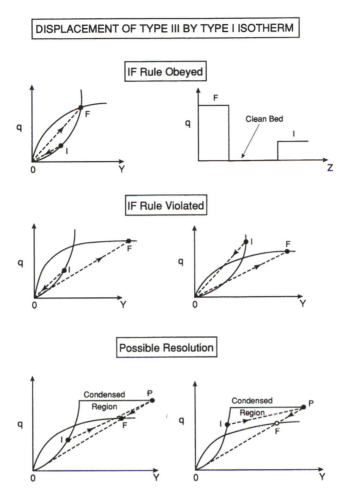

FIGURE 3.6
Displacement of solute from a type III isotherm by various concentration levels of a type I isotherm.

Resolution of the dilemma hinges on the recognition that type III isotherms ultimately reach a condensed region, or else veer off to become type V isotherms (see the case of carbon moisture isotherms, Figure 1.3). In either case, new pathways are opened up which lead *upward* and away from the origin. We have sketched these in the third case shown in Figure 3.6. In both cases a plateau P is attained which lies in the condensed region, representing a considerable enrichment over the original level of I. Similar cases were encountered in cases E and F of the preceding illustration. We caution the reader again that the rear zone descending from I via F to the origin may on occasion be an expanding one. This does not, however, alter the fact that a considerable enrichment from I to a plateau P is attained.

Chapter 4

EQUILIBRIUM EFFECTS: ADIABATIC SORPTION

The intrusion of temperature brings about an escalation in complexity similar to that which accompanied the entry of a second sorbable component: we are now effectively dealing with a two-component, or two-variable, system. Fortunately the pain is lessened by the fact that graphical analysis can still be carried out on a single q-Y diagram, with temperature T as a floating parameter. This approach can in some cases be translated into simple algebraic expressions.

Temperature effects are negligible in most liquid phase applications but almost always make themselves felt in gaseous systems. In adsorption, the temperature rise may amount to no more than 1 to 2°C, in which case the assumption of isothermal operation will not severely distort the prediction. However, ΔTs of several ten degrees are not uncommon. These can no longer be dismissed since they significantly reduce the capacity of the bed and alter the course of sorption. In desorption, which is often carried out with a hot purge gas, the effects are even more pronounced and the inclusion of the temperature variable becomes mandatory.

The major parameters we encountered in isothermal sorption, e.g., breakthrough and desorption times, or minimum bed and purge requirements, will retain their importance but, in addition, we will be addressing questions peculiar to nonisothermal systems, e.g.,

> What is the magnitude of the temperature rise? How is it affected by system parameters, for example pressure?

In what fashion do temperature and concentration fronts
interact? What are the profile shapes?

What degree of enrichment can one expect during hot
purge regeneration?

Other subsidiary questions will be dealt with as they arise
in the context of major questions. In the discussion which follows
we limit ourselves initially to answering some simple but important questions, leaving the more complex issues of profile shape,
etc., to a later section. It should be noted that all topics will be
based on the assumption of local equilibrium, as was the case in
our preceding treatment of isothermal systems.

4.1   MAGNITUDE OF THE TEMPERATURE RISE ΔT

The first question to be addressed is whether we should
deviate from isothermal analysis. In desorption of gases the
answer is invariably yes, since temperature changes are usually
in excess of 1–2 °C. In adsorption the answer is not as straightforward, but a simple energy balance under equilibrium conditions provides an accurate estimate of the maximum temperature
rise to be expected.

$$\Delta T = T_{Max} - T_F = \frac{q\Delta H/C_{pb}}{(q/Y)_F - C_{ps}/C_{pb}} \qquad (4.1)$$

where the subscript F refers to inlet (feed) conditions, ΔH is the
heat of adsorption, and C_{ps}, C_{pb} are heat capacities of sorbent
and carrier gas, respectively. Since $C_{ps}/C_{pb} \ll (q/Y)$, under most
practical conditions, this reduces to the simple expression

$$\Delta T = Y_F\Delta H/C_{pb} \qquad (4.2)$$

which can be used for quick estimates of the maximum temperature rise. For example, with ΔH having a range of 1000–4000
kJ/kg for most sorbates (average 2500), C_{pb} being of the order 1
kJ/kg K for common carrier gases and Y_1 set at a typical inlet
concentration of 1% (0.01 kg/kg carrier) one obtains

$$\Delta T = 0.01 \cdot 2500/1 = 25\,°C$$

Temperature rises of 25°C are thus not uncommon in the adsorption of gaseous systems. There are several further deductions to be made from this simple expression:

- Since at constant partial pressure Y_F is inversely proportional to total pressure, ΔT will decrease with an increase in total pressure. Typically isothermal conditions are then obtained at $P_{tot} \cong 50$ atm.
- Carrier gases with high specific heats (e.g., hydrogen) will similarly tend to suppress the temperature rise.
- ΔT increases in direct proportion to inlet concentration. This is expected on physical grounds but proof of the linear relation requires recourse to the energy balance given above.

4.2 QUALITATIVE INTERACTION OF TEMPERATURE AND CONCENTRATION FRONTS

We start by presenting the two major types of (T, Y, q) fronts which arise in adiabatic sorption. They are termed, respectively, and for reasons which will become apparent, pure thermal wave front (PTF) and combined wave front (CWF). Their properties and to some degree their practical significance are summarized below, and displayed in Figure 4.1.

4.2.1 Pure Thermal Wave Front (PTF)

- In this case a pure temperature front, devoid of any solute, detaches itself from the main concentration front and moves ahead of it (adsorption) or lags behind it (desorption). This results in a pure temperature plateau, or clean bed, opening up between the main concentration and temperature fronts in much the same way as happened in the isothermal displacement with feed below the watershed point.
- PTFs are the most common fronts encountered in practical adsorption operations. In these processes it is usual to

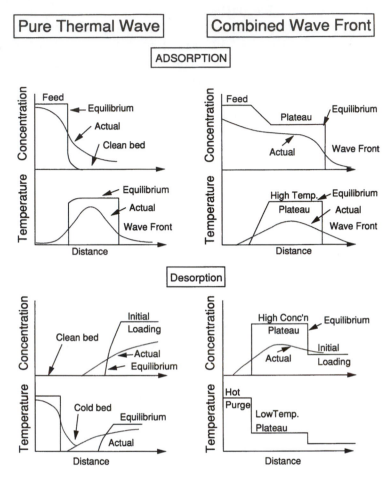

FIGURE 4.1

Temperature effects in sorption: the two types of waves that arise in adiabatic adsorption and desorption.

encounter temperature waves running ahead of the concentration front. The temperature rise associated with the wave is given by the previously cited relations (4.1) and (4.2), which are valid for PTFs only.

- In contrast, PTFs are rarely encountered in practical *desorption* operations. In other words, one does not see, except under unusual circumstances, a pure temperature wave following a front of stripped solute.

4.2.2 Combined Wave Front (CWF)

- In this type of front, both temperature and concentration move in unison along the entire profile. No separation between the two occurs. Temperature plateaus do make an appearance but are contaminated with solute, i.e., do not lead to the clean bed seen before. This parallels the isothermal case when displacement is carried out with feed *above* the watershed point and leads to combined propagation of the two solutes.

- CWFs are most commonly encountered in practical *desorption* operations. In these operations one sees a joint propagation of desorbed solute and temperatures. In contrast, it is unusual to encounter CWFs in practical *adsorption* operations, since this requires unusually high feed concentrations. The reader is reminded here that adsorption is usually applied to feed concentrations of 1–2% or less, higher levels leading to inefficient or costly operation. Cost and inefficiency are a direct result of the formation of a CWF, which leads to a warming up of the bed, thus reducing capacity. In PTF, by contrast, much of the heat is propagated and removed *ahead* of the solute front, thus leading to higher sorptive capacity of the bed.

Two criteria, both involving inequalities, are available for delineating the operating conditions which give rise to the two types of fronts. They are

$$
\left(q/F\right)_F \underset{>}{\overset{<}{}} C_{ps}/C_p \quad \text{and} \quad \frac{q\left(T_{max}, Y_F\right)}{Y_F} \underset{>}{\overset{<}{}} C_{ps}/C_{pb}
$$

Both inequalities hinge on whether the feed ratio $(q/Y)_F$ is greater or smaller than the specific heat ratio of solid to carrier gas. These are the preeminent parameters which determine the course of adiabatic sorption. Their combined effect is summarized in the Tables below.

For desorption, less comprehensive criteria exist which involve the loading ratio $(q/Y)_I$ at the initial bed conditions rather than the feed ratio $(q/Y)_F$. We display them below.

TABLE 4.1

Criteria for PTF/CWF Formation in Adsorption

	$\dfrac{q\left(T_{max}, Y_F\right)}{Y_F} > C_{ps}/C_{pb}$	$q(T_{Max}, Y_F) < C_{ps}/C_{pb}$
$(q/Y)_F > C_{ps}/C_{pb}$	Only PTF formed	Either PTF or CWF formed
$(q/Y)_F < C_{ps}/C_{pb}$	Only CWF formed	Only CWF formed

TABLE 4.2

Criteria for PTF/CWF Formation in Desorption

$(q/Y)_I > C_{ps}/C_{pb}$	Only CWF formed
$(q/Y)_I < C_{ps}/C_{pb}$	CWF or PTF formed

Examples of PTFs and CWFs for both adsorption and desorption appear in the accompanying diagram (Figure 4.1). These important waves have a number of additional features of great practical importance, which we summarize below.

- The principal feature of adsorption PTFs is that the heat generated in the process is propelled ahead of the main solute front, thus allowing adsorption to proceed efficiently on a "cold" bed near ambient temperatures. This propagation sequence is not limited to heat effects produced by the adsorption process. Thus, if the bed is *initially hot*, because of a previous hot purge regeneration process, that stored heat will similarly be swept ahead of the solute front, which will then encounter a bed which has essentially been cooled to ambient conditions. The important corollary is that breakthrough times will be approximately the same, irrespective of whether the feed is introduced to a hot or cold bed. Similarly, minimum bed requirements will stay the same regardless of the initial bed temperatures, thus avoiding the necessity to cool the bed. This important feature has been exploited in a number of practical processes.

- PTFs in desorption, though much rarer, lead to a similar separation of solute and thermal fronts, which, however, propagate in reverse order, solute preceding temperature. One sees, or surmises, that the wave of stripped solute proceeds quite independently of the temperature wave that follows it. The important practical implication in this behavior is that the entire desorption process becomes independent of the purge gas temperature, since the solute front is moving independently on the subsequent thermal wave. It follows in turn that desorption time and minimum purge requirements will remain the same whether desorption is carried out with a hot or cold gas. The reader is reminded, however, that these conditions materialize only if the criteria presented in the preceding table are satisfied, i.e., at the very least we must have $(q/Y)_I < C_{ps}/C_{pb}$.

- The negative features of CWFs in adsorption have already been mentioned: high plateau temperatures which reduce bed capacity and render the process inefficient. In desorption the reverse is true. Here we see the emergence of high *concentration* plateaus accompanied by relatively low temperatures (see Figure 4.3). The consequences are not trivial. A high solute plateau which exceeds the initial loading concentration carries with it the immense benefit of *enriching* adsorbed solute above its feed concentration. Thus, we achieve, at one and the same time, removal as well as enrichment of the absorbed component, making it easier to achieve final recovery, for example by condensation. We shall shortly establish simple upper limits to the degree of enrichment to be expected.

4.3   REPRESENTATION IN THE q-Y DIAGRAM

Trajectories of adiabatic sorption can be traced out in a q-Y diagram much as was done in the case of binary isothermal Langmuir sorption. Two sets of characteristics result, which, as before, we denote by q^+ and q^- (Figure 4.2). They arise from a set of differential equations whose form or solution need not concern us here, since we shall shortly present a graphical procedure for the construction of q^+ and q^- from simple system parameters.

We shall limit ourselves to the trajectories of the two cases which arise most frequently:

- Adiabatic adsorption on a clean bed at low concentration which we have seen leads to a single solute front, hence requires only a single characteristic, a q^- on which the feed point is located. The second characteristic in this case

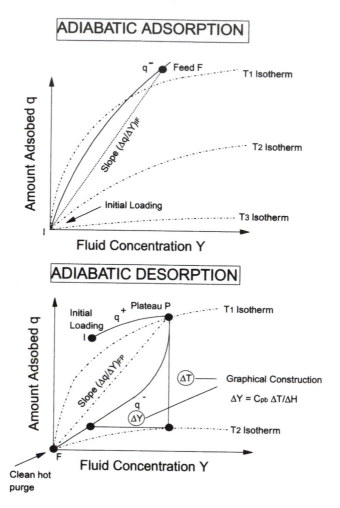

FIGURE 4.2

Characteristics (pathways) in adiabatic sorption. Feed F is located on a negative characteristic q⁻, the initial bed condition I on a positive characteristic q⁺. Construction of the characteristics proceeds via simple algebraic algorithms, Equations (4.3) or (4.4).

> resides at the origin and comprises the pure temperature front.

- Adiabatic desorption which most often leads to two combined fronts separated by a plateau and hence gives rise to two branches along q⁺ and q⁻.

These cases have been sketched in the accompanying diagrams. We summarize below the main features and procedures for using and exploiting these figures.

- Both Rule 1 (IF Rule) and Rule 2 of isothermal sorption apply here as well. Thus, the initial loading point, I, is located on a q^+, and the feed point, F, on a q^-. In proceeding from I to F, the slope of the trajectory must either stay constant (yielding a discontinuous front) or increase (leading to an expanding front). When the trajectory involves inflections, the procedures established for inflecting isotherms are invoked.
- The q^+ and q^- characteristics cross various equilibrium isotherms whose intersection identifies the prevailing temperature at that point.
- Intersection of a q^+ and q^- signals the formation of solute and temperature plateaus which increase in length or duration as sorption progresses.
- Construction of q^+ and q^- which constitute the trajectory can make use of the following simple relation:

$$\Delta Y = \frac{C_{pb}\Delta T}{\Delta H} \tag{4.3}$$

Construction typically starts at the feed point F or the initial loading point I, located at T_1, Y_1. A convenient isotherm interval $T_1 - T_2 = \Delta T$ is chosen, say 10 or 20 °C, and multiplied by the ratio of carrier heat capacity to heat of adsorption $C_{pb}/\Delta H$. This ratio, as we have seen, is typically of the magnitude 1/2500. The result yields ΔY, or the location of Y_2 on the T_2 isotherm. This is continued repeatedly until intersection of the two branches or the origin occurs.

The procedure is valid for values of $\dfrac{\Delta H \cdot \Delta q}{C_{pb}\Delta T} \gg C_{ps}/C_p$, i.e.,

when solid sensible heating is small compared to latent heat effects. When this is not the case, the following full equation is applied:

$$\frac{\Delta q}{\Delta Y} = \frac{C_{ps}}{C_{pb}} + \frac{\Delta H \cdot \Delta q}{C_{pb}\Delta T} \tag{4.4}$$

The procedure here consists of choosing a temperature interval ΔT between isotherms, assuming a value for Δq and adjusting it together with ΔY until Equation (4.4) is satisfied.

4.4   DESIGN EQUATIONS, MINIMUM BED AND PURGE REQUIREMENTS

These are established in the same way as for isothermal sorption, using the appropriate q^+ or q^- as an "effective" equilibrium curve of type I or type III along which adsorption or desorption is taking place. Thus, for adsorption along a q^- which is concave upward, we declare the process to be taking place along an effective type III isotherm or, if concave downward, along a favorable type I isotherm. For desorption, we usually neglect the purge gas consumption incurred along the shallow branch, and base our calculations on events along the steep branch where the overwhelming majority of consumption occurs. This branch is usually of type III and desorption along it leads to a discontinuous front, as it would in isothermal desorption. When inflections occur, we resort to the appropriate procedure applicable to the isothermal case. A summary of the pertinent relations appears below.

1. Adsorption: q^- concave upward

Design Equation:
$$V = z/t = \frac{G_b}{\rho_b \cdot H} \tag{4.5}$$

Breakthrough Time:
$$t_{ads} = \frac{\rho_b \cdot H}{G_b} \times \text{length of bed L} \tag{4.6}$$

Minimum Bed Requirement: $W_m = 1/H$ kg bed/kg carrier   (4.7)

2. Adsorption: q^- concave downward

Design Equation:
$$V = z/t = \frac{G_b}{\rho_b \cdot (\Delta q/\Delta Y)_{IF}} \tag{4.8}$$

Breakthrough Time:
$$t_{ads} = \frac{\rho_b \cdot (\Delta q/\Delta Y)_{IF}}{G_b} \times \text{length of bed L} \tag{4.9}$$

Minimum Bed Requirement: $W_m = (\Delta Y/\Delta q)_{IF}$ kg bed/kg carrier   (4.10)

3. Desorption: q⁻ concave upward

Design Equation:

$$V = z/t = \frac{G_b}{\rho_b \cdot (\Delta q/\Delta Y)_{FP}}$$

(4.11)

Desorption Time:

$$t_{des} = \rho_b \frac{(\Delta q/\Delta Y)_{FP}}{G_b} \times \text{length of bed L}$$

(4.12)

Minimum Purge
Requirement:

$$G_m = \Delta q/\Delta Y \text{ kg purge/kg bed}$$

(4.13)

4.5 MAXIMUM ENRICHMENT ATTAINABLE IN HOT PURGE DESORPTION; STEAM REGENERATION

These important aspects are established by neglecting sensible heat effects (the heat required to raise the temperature of the bed) and applying Equation (4.3) to both branches of the desorption trajectory. Upon eliminating the plateau temperature T_p from the two equations one obtains for the plateau concentration Y_p

$$\left(Y_p\right)_{Max} = \frac{C_{pb}}{2\Delta H}\left(T_R - T_F\right) + \frac{Y_F}{2}$$

(4.14)

where T_R and T_F are the regenerant and adsorption feed temperatures, respectively. Y_p represents a *maximum* value because of the omission of the sensible heating of the solid. Thus, for feed conditions $Y_F = 0.01$, $T_F = 25\,°C$, and regenerant temperatures of

$$300\,°C, (Y_p)_{Max} \cong \frac{1}{5000} 275 + \frac{0.01}{2} = 0.06 \text{ , a sixfold enrichment.}$$

Four important subcases result from this treatment:

1. If the regeneration temperature is set equal to that of the adsorber feed, the plateau concentration will *at most* equal *one half* the adsorber feed concentration:

$$Y_p = \frac{Y_F}{2}$$

(4.15)

2. The minimum regenerant temperature required to *exceed* Y_F is given by

$$T_R = \frac{\Delta H Y_F}{C_p} + T_F \qquad (4.16)$$

Thus, for a typical value of $\Delta H/C_p = 2500$ and $Y_F = 0.01$, purge temperature T_R must exceed that of the adsorber feed by at least $>25\,°C$ in order to achieve enrichment in the feed concentration.

3. If the adiabatic desorption curve is known or estimated (without knowing precisely the corresponding purge temperature), then the enrichment Y_p/Y_F is given by the following approximate relation (see Figure 4.2)

$$\frac{Y_p}{Y_F} \cong \frac{\text{Slope of adsorption operating line}}{\text{Slope of desorption operating line}} = \frac{\text{Slope of FI}}{\text{Slope of FP}} \qquad (4.17)$$

This merely confirms the self-evident fact that the further to the right one positions the steep branch of the desorption curve, the higher the plateau value Y_p.

4. For steam regeneration, the latent heat of water ΔH_{H_2O} becomes the driving force and leads to the approximate relation

$$\left(Y_p\right) \cong \frac{\Delta H_{H_2O}}{\Delta H} = 0(1) \qquad (4.18)$$

Since both ΔH_{H_2O} and the "typical" enthalpy of adsorption are of similar magnitude (approximately 2500 kJ/kg) peak mass ratios $(Y_p)_{Max}$ will be of the order of unity. As expected this is considerably higher than the Y_p values obtained by hot gas regeneration. It is reasonable to assume that this will lead to the formation of a separate phase of condensed solute in contact with a phase of saturated steam condensate. This mode of regeneration is generally restricted to the recovery of adsorbed vapors from carbon beds; both silica gel and zeolites decompose in contact with steam.

To further illustrate the effect of regeneration temperature, we display in Figure 4.3 desorption breakthrough curves obtained at two different purge temperatures. Both display plateaus flanked by transient zones. When purge is carried out at ambient temperature, the fluid phase concentration and purge temperature quickly drop to low plateau values which persist for a considerable period. Desorption is clearly an inefficient process here and no enrichment in the desorbing fluid is obtained. At high purge temperature, two effects are noticed: the desorption time is shortened and, most importantly, the fluid concentration rises above the initial loading value, thus providing valuable enrichment in the desorbing component. The bulk of the desorption process is achieved at a surprisingly low temperature (the intermediate temperature plateau). This is because the sensible regenerant temperature is used to provide the necessary latent heat of desorption.

Illustration 4.1: Transition from PTF to CWF — Analysis of Air Drying Operations

The thermal behavior in adiabatic adsorption is crucial to the efficiency of the process. As we have seen, when a pure thermal wave (PTF) is formed, much of the heat is carried out ahead of the sorption front and adsorption can proceed efficiently at low or near ambient temperatures. The formation of a combined wave front implies that the sorption front is raised to the adiabatic temperature, resulting in a drastically reduced bed capacity. We analyze this behavior in the context of one of the most common adsorption operations, the drying of air.

PTF Formation — The diagram given for drying operations, Figure 4.4, shows a single curve connecting feed and initial bed condition at the origin. Thus, pure thermal wave formation occurs for all three major commercial adsorbents, even at 100% relative humidity, provided the feed is introduced at 25 °C. The criterion to be satisfied is that the feed concentration ratio $(q/Y)_F$ has to exceed the heat capacity ratio C_{ps}/C_{pb}, which is approximately 1 for the systems under consideration. Results for the three major commercial drying agents are shown below for both 25 °C and 60 °C, 100 kPa.

It is clear that saturated moist air at 25 °C is capable of producing a pure thermal wave and is thus a good candidate for

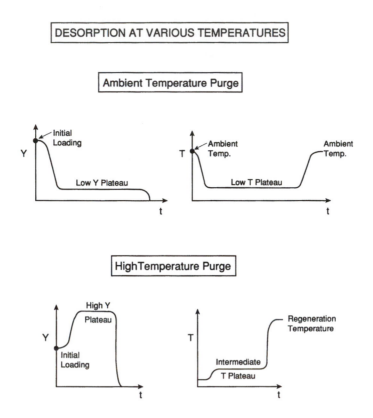

FIGURE 4.3
Desorption breakthrough curves at different purge temperatures. Hot purge leads to solute enrichment, ambient purge to levels below initial loading.

adsorptive drying. At lower relative humidities or higher total pressures, conditions are even more favorable. High-pressure operation, in particular, is a preferred mode of operation, since it brings the system even closer to isothermal conditions (see previous illustration on inflecting isotherms).

CWF formation — This occurs when air temperature is raised to 60 °C at full saturation. Moisture content is now at approximately 0.13 kg H_2O/kg air. Saturation capacities q for alumina and silica gel are at ~0.1 kg H_2O/kg solid (beyond the range of Figure 4.4). Both alumina and silica gel are clearly below the required criterion of $(q/Y)_F$ >1, and the zeolite, while still in principle capable of producing pure thermal waves, is a marginal case. Thus, for these moisture concentrations, which are approximately ten times higher than in the previous case, adsorptive drying at

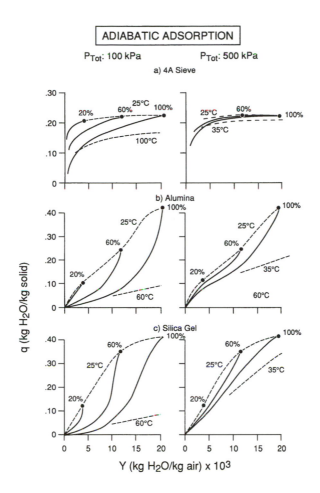

FIGURE 4.4

Adiabatic adsorption on various commercial desiccants. Heavy lines represent the characteristic pathways connecting feed air of various humidities to the clean bed located at the origin. Broken lines are equilibrium isotherms. (Reproduced from Basmadjian, D., *Advances in Drying*, Vol. 3, p. 326. Taylor & Francis, Washington, D.C., 1984. Reproduced with permission. All rights reserved.)

	Zeolite	Alumina	Silica Gel
$(q/Y)_F$ 25°C, 100% RH	11	21	21
$(q/Y)_F$ 60°C, 100% RH	1.4	0.77	0.77

atmospheric pressure is clearly to be rejected. Higher operating pressures, however, would tend to remedy the situation and at 50 atm one can expect isothermal conditions to prevail.

Illustration 4.2: Desorption by Hot Purge

We have seen that hot purge desorption leads to a two-branch trajectory from the initial loading point to the feed (purge gas represented by the origin in Figure 4.2). One branch, a very shallow, nearly horizontal one, contributes little to the desorption process, and serves instead to raise the bed to an intermediate plateau temperature reminiscent of the wet-bulb temperature familiar from evaporation processes. Nearly all of the desorption takes place along the steep branch which descends from the plateau intersection to the feed point, i.e., the origin. It is the characteristics of this branch which determine the operational parameters such as desorption time and purge consumption.

We choose one of the cases described in the accompanying diagram, Figure 4.5, the desorption of a bed of silica gel saturated at 25°C and 60% RH at 500 kPa with a purge temperature of 150°C. The corresponding adsorption diagram shows that operation to be nearly isothermal, so that the assumption of a uniform temperature at the start of the operation is justified. We set velocity at $v = 1$ m/s and bed height at $z = 1$ m. Heat capacities can be taken as unity (kJ/kg K), and the density ratio rounded off to $\rho_b/\rho_f = 10^3$.

Plateau temperature T_p — This parameter would have been established during the construction of the adiabatic equilibrium curves but was not entered in the diagram. We make use instead of the same algorithm which was used in the construction to obtain an analytical expression for T_p. We have

$$\frac{\Delta q}{\Delta Y} = \frac{C_{ps}}{C_{pb}} + \frac{\Delta T}{C_{pb}\Delta T}\Delta q$$

where $\Delta H = 3100$ kJ/kg for water on silica gel. The first term on the right and the differences refer to those along the steep branch side and can be neglected in comparison to the latent heat term. We obtain

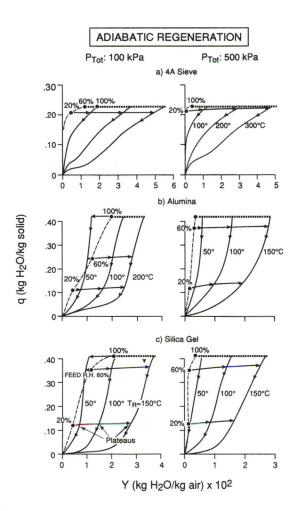

FIGURE 4.5

Adiabatic regeneration by hot gas purge of various commercial desiccants. Heavy lines, consisting of a shallow and steep branch, represent character- istics which intersect at a plateau. Broken lines are isotherms, dotted lined the condensation boundaries. T_R, temperature of hot gas regenerant. (Re- produced from Basmadjian, D., *Advances in Drying,* Vol. 3, p. 337. Taylor & Francis, Washington, D.C., 1984. Reproduced with permission. All rights reserved.)

$$\Delta T = \Delta Y \cdot \Delta H / C_{pb} = 0.025 \cdot 3100 / 1 = 81\,^\circ C$$

Hence, $T_p = 150 - 81 = 69\,^\circ C$.

Operating line — That line is taken to be the chord from the origin (purge) to the plateau and has a slope $\Delta q / \Delta Y = 0.36/2.5 \cdot 10^{-2} = 17.4$.

Desorption time — We employ the customary relation

$$t_{des} = \rho_b \frac{\Delta q / \Delta Y}{G_b} \cdot L = \frac{\rho_b}{\rho_f} \cdot \frac{(\Delta q / \Delta Y)L}{v}$$

and obtain

$$t_{des} = 10^3 \cdot 17.4 \cdot 1/1 = 1.74 \cdot 10^4 \text{ s}$$

Minimum purge requirement — This parameter equals the slope of the operating line and we have

$$G_m = \Delta q / \Delta Y = 17.4 \text{ kg purge/kg bed}$$

Desorption of zeolite beds — It is of some interest to compare the performance of silica gel with that of a zeolite bed. The prospects for the latter appear dim, since the effective equilibrium curves are all wholly or in part of type I (see Figure 4.5), so that desorption will be dependent on Henry's constant, which is large compared to slopes of operating lines we have seen. Indeed, at a purge temperature of 100 °C, we estimate it to be in the range 100–200 kg purge/kg bed so that gas consumption is an order of magnitude larger than for the silica gel bed purged at 150 °C. However, when purge temperature is raised to 300 °C, Henry's constant drops to more reasonable levels, $H \cong 10$, so that gas consumption is roughly on a par with that of a silica gel bed. However, this improved performance is brought about at the cost of a higher purge gas temperature (300 °C vs. 150 °C).

Illustration 4.3: A Criterion for the Entire Cycle

Purge requirements can also be based on the amount of gas treated, which is simply done by forming the product of minimum bed and purge requirements:

Relative purge consumption = R [kg purge/kg carrier treated]
　　　 = W_m [kg bed/kg carrier] G_m [kg purge/kg bed]

This has the advantage of providing a single figure for the entire cycle of this major expense, which together with pumping comprises most of the total operating cost. It also provides us with a convenient efficiency criterion, for when R = 1, as much gas would be required for purge as has been treated or produced. Typically one should therefore aim at R values of 10% or less. This can often be achieved by operating at elevated pressures.

We briefly consider the case of the silica gel bed given in the previous illustration. The operating diagram for the adsorption step is shown in Figure 4.4 and yields a slope H_{ads} = 0.35/2.5 · 10^{-3} = 140 kg air/kg bed. For desorption, we had $(\Delta q/\Delta Y)_{des}$ = 17.4. Hence,

$$R = \frac{1}{H_{ads}} \cdot (\Delta q/\Delta Y)_{des} = \frac{1}{140} \cdot 17.4 = 0.12$$

Thus, 12% of purified gas is required as purge.

To ease the computations, we have displayed minimum bed and purge requirements for drying of air in the accompanying diagrams, Figure 4.6 and 4.7.

Illustration 4.4: Drying of a Steam Regenerated Carbon Bed — Troubleshooting

Carbon beds used in solvent recovery are often regenerated (desorbed) with steam, followed by a purge with air to drive off condensed moisture. One such unit had been used in a plant to remove vinyl chloride vapor, a known carcinogen, from plant air. The carbon bed, 1.2 m in diameter and packed to a height of 1.4 m, had performed satisfactorily in single test runs (one cycle) but quickly broke down on repeated cycling. Inspection of the bed showed considerable residual moisture on the carbon particles. The focus was thus quickly shifted to inefficiencies in the drying step.

The plant operations had provided a blower and heater which supplied air at approximately 0.1 m³/s ($\cong$ 0.1 kg/s) and at a temperature of 50°C to the carbon bed, which was assumed to be at a temperature of about 100°C from the previous regeneration step. One hour to $1^1/_2$ hours was allowed for the drying step.

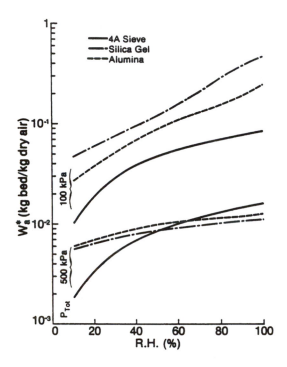

BED REQUIREMENTS FOR DRYING OF AIR

FIGURE 4.6
Minimum bed requirements in the drying of air with various desiccants.
(Reproduced from Basmadjian, D., *Advances in Drying*, Vol. 3, p. 329. Taylor
& Francis, Washington, D.C., 1984. Reproduced with permission. All rights
reserved.)

Part of the successful remedy hinged on the realization that
the problem had little to do with adsorption per se: most of the
steam condensate is present as free moisture or a residual mist
in the bed voids. It was further realized that this free moisture
would quickly drop in temperature to "wet-bulb" or adiabatic
saturation conditions with an evaporation rate vastly reduced
over that prevailing at 100°C.

A good approach is to establish *minimum* conditions, which
will dry the bed in the prescribed time, much in line with the
approach used in equilibrium theory. It has previously been

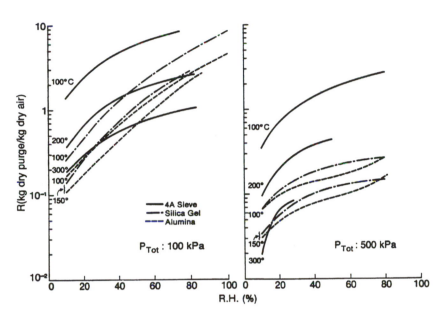

PURGE REQUIREMENTS
FOR DRYING OF AIR

FIGURE 4.7
Minimum purge requirements in the regeneration of desiccants. (Reproduced from Basmadjian, D., *Advances in Drying,* Vol. 3, p. 330. Taylor & Francis, Washington, D.C., 1984. Reproduced with permission. All rights reserved.)

estimated that approximately 100 kg of free moisture resided in the bed. Such rough estimates are often arrived out by summing latent and sensible heat requirements, including the heat used to bring the vessel wall to the required temperature. Theoretical treatments of this problem are sparse and incomplete and one must perforce fall back on the cruder approach outlined above.

To establish the minimum conditions for flow rate and incoming air temperature, we assume that the air leaves at equilibrium, i.e., at the prevailing adiabatic saturation temperature and corresponding moisture content Y_{as}.

The total amount of moisture carried away by the saturated air must equal the total to be evaporated. Thus,

$$Y_{as}\left(\frac{kg\ H_2O}{kg\ air}\right)\cdot G\left(\frac{kg\ air}{s}\right)\cdot t = 100\ kg\ H_2O$$

or

$$G = \frac{100}{Y_{as}\cdot 5400} = \frac{1}{54\ Y_{as}}$$

We have used standard psychrometric charts to establish Y_{as} and tabulate the results below

T, °C	25	50	75	100	125
$Y_{as}\left(\dfrac{kg\ H_2O}{kg\ air}\right)$	0.006	0.013	0.021	0.028	0.038
G (kg/s)	3.1	1.4	0.88	0.66	0.49

The means provided for, G = 0.1 m³/s at 50°C, are clearly inadequate. Recommendations must therefore be made for both higher inlet temperatures and flow rates. The reader is warned that remedies calling for extensive changes in equipment are unpalatable to plant managers. The example also serves to show that principles from outside the immediate absorption field must sometimes, in fact quite frequently, be called upon to result in the successful application of adsorption operations.

Chapter 5

EQUILIBRIUM EFFECTS: MULTICOMPONENT SORPTION

This is a formidable subject, ordinarily handled by elaborate computer simulations or near-incomprehensible theory. To all appearances it has become the preserve of academic research, while the outside world continues to treat multicomponent systems by rule of thumb, extensive experimentation, or in-house computer simulations. The gap between the two approaches is a wide one.

Computer simulations have largely been limited to two- or three-component systems, being hampered by a lack of suitable equilibrium data or multicomponent prediction methods. Rules of thumb and experimentation are limited to specific systems which occasionally allow extrapolation to more general rules. Academic research, while very precise in its approach, has been mainly aimed at model validation. This is usually accomplished by showing agreement between experimental and predicted breakthrough curves, requiring input of independent equilibrium data and transport rate coefficients. The larger issues of analyzing these systems or obtaining parameters of direct interest to the practitioner are not addressed. We propose here that a few things can be done to remedy this problem by invoking the simple principles we have laid out for one- and two-component systems. The answers most often constitute upper or lower limits to actual behavior. Nevertheless they provide useful guidelines and often allow deeper insight to be gained into the behavior of these systems.

5.1   LINEAR ISOTHERMAL SYSTEMS

Here we consider mixtures at the trace level, of which each component obeys Henry's law. Some fairly precise answers can be given in this case.

Each component propagates independently of the others, resulting in a sequence of discontinuous and sharp fronts during both adsorption and desorption. We are thus able to focus on the leading (or "light") component 1 to derive parameters of interest. Following what has been said for single components, we obtain the following relations

Design Equation:

$$v = z/t = \frac{G_b}{\rho_b \cdot H_1} \tag{5.1}$$

Breakthrough Time:

$$t_{abs} = \frac{\rho_b \cdot H_1}{G_b} \times \text{length of bed L} \tag{5.2}$$

Minimum Bed Requirement:

$$W_m = 1/H_1 \text{ kg bed/kg carrier} \tag{5.3}$$

where G_b can also be expressed as $\rho_f \cdot v$. We demonstrate the use and extension of the equations by means of the following.

Illustration 5.1: Sorption of a Mixture of Aqueous Pollutants on Activated Carbon (Linear Case)

We consider the range $H = 10^3 - 10^4$, which applies to the more stubbornly held pollutants and use a bed of 1 m and a fluid velocity of 10^{-3} m/s. The light component with $H = 10^3$ will break through first and determines bed requirements. The same system with transport resistance is taken up in a later section (Illustration 8.4).

Breakthrough time — This is immediately given by the customary relation

$$t_{ads} = \left(\frac{\rho_b}{\rho_f}\right)\frac{H_1}{v} = \frac{440}{1000} \cdot \frac{10^3}{10^{-3}} = 4.4 \cdot 10^5 \text{s}$$

Minimum bed requirement — Drawing on Equation (5.3) and applying it to the light component we obtain

$$W_m = 1/H_1 = 1/10^3 = 10^{-3} \text{ kg carbon/kg water}$$

Fraction of bed occupied by heavy component — Here we use the "design" equation and arrive at the following value

$$z = v \cdot t \cdot \frac{\rho_f}{\rho_b} \cdot \frac{1}{H_3} = 10^{-3} \cdot 4.4 \cdot 10^5 \cdot \frac{1000}{440} \cdot \frac{1}{10^4} = 0.1 \text{ m}$$

In other words, the heavy component takes up 10% of the bed at breakthrough, as would have been expected from the ratio of Henry constants.

5.2 NONLINEAR ISOTHERMAL SYSTEMS

This complex topic is now the exclusive domain of elaborate theories, computer simulations, and elaborate experimentation. Still, one or two things can be done to alleviate the pain. To begin with, we wish to convey a sense of the physical behavior of such systems, which we do in Figure 5.1 for a set of Langmuir solutes. The profiles show the light component preceding the heavier components, as expected from the corresponding binary case. The novelty comes from the interjection of the intermediate component, which gives rise to mixed plateaus flanked by mixed bands of components. In all other respects, however, the qualitative features are those of the binary case.

- In adsorption the light component breaks through ahead of all others in *pure* form and at a plateau level which exceeds the feed concentration. The heavy component breaks through last, rising directly to its feed concentration without exceeding it. The intermediate component also rises above its feed concentration.

- In desorption the light component again comes off ahead of all others, descending to zero concentration in a single expanding front. The heavy component desorbs last after forming a pure plateau devoid of light and intermediate component. Mixed plateaus occur in all other positions.

The fact that the heavy component descends in pure form to zero concentration allows us to make some precise calculations

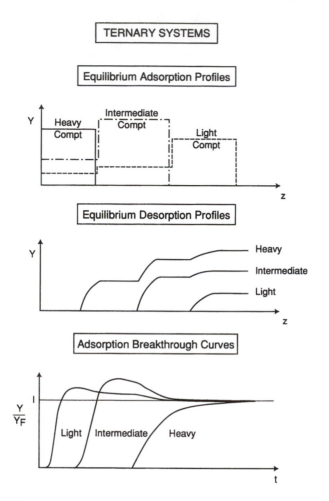

FIGURE 5.1
Profiles and breakthrough curves for ternary Langmuir systems.

regarding the desorption process, even in this nonlinear case. We demonstrate this in the following illustration.

Illustration 5.2: Purge of Nonlinear Solutes

We consider the purge of a multicomponent Langmuir-type mixture of aqueous pollutants from a carbon bed. The isotherms are not known (they rarely are), but Henry constants are estimated at $H = 6.6 - 500$ kg water/kg carbon (see Table 7.1). We set v = 10^{-2} m/s, z = 1 m. Carbon bed density ρ_b = 450 kg/m^3. Purge is determined by the highest Henry constant. We obtain

Purge Time:

$$t_{des} = \frac{\rho_b}{\rho_f} \cdot \frac{H}{v} L = \frac{450}{1000} \frac{500}{10^{-2}} 1 = 2.25 \cdot 10^4 \text{s}$$

Minimum Purge Requirement:

$$G_m = H = 500 \text{ kg water/kg carbon}$$

The exorbitant amount of purge required attests to the long, drawn out nature of the desorption process occasioned by the high Henry constant.

No such precise predictions can be made for the adsorption case. We use instead the approach of bracketing the solution. This is done by recognizing that the operating line for the plateau of the light component which breaks through first must lie between its Henry constant and the slope of the line connecting feed of component 1 to the origin. We are able to write:

$$H_1 \quad > \quad \Delta q_1 / \Delta Y_1 \quad > \quad (q/Y)_1$$
Henry constant   Operating line   Feed point

The only equilibrium point needed for this approach is the loading of the light component 1 at its feed composition Y1. Thus, there is no need for a complete set of multicomponent equilibria to be determined. We illustrate this by means of the following example.

Illustration 5.3: Sorption of a Mixture of Aqueous Pollutants on Activated Carbon (Nonlinear)

We retain the same parameters as in the linear case, i.e., z = 1 m, v = 10^{-3} m/s. Feed concentration of the light component is set at $4.7 \cdot 10^{-4}$ kg/kg water, with an equilibrium loading of 0.05 kg/kg carbon. The Henry constant is maintained at $H_1 = 10^3$ kg water/kg carbon. We obtain

$$H_1 = 10^3 > \Delta q_1/\Delta Y_1 > (q/Y)_1 = \frac{0.05}{4.7 \cdot 10^{-4}} = 1.06 \cdot 10^2$$

Breakthrough time — We bracket t_{ads} between the two values given above by setting

$$t_{abs} = \frac{\rho_b}{\rho_f} \cdot \frac{H_1}{v} L \quad \text{and} \quad t_{abs} = \frac{\rho_b}{\rho_f} \cdot \frac{(q/Y)_1}{v} L$$

Thus,

$$\frac{440}{1000} \cdot \frac{10^3}{10^{-3}} 1 > t_{abs} > \frac{440}{1000} \cdot \frac{1.06 \cdot 10^2}{10^{-3}} 1$$

and hence

$$4.4 \cdot 10^5 > t_{ads} > 4.7 \cdot 10^4$$

Minimum bed requirement — The calculation proceeds in similar fashion as for breakthrough time

$$1/H_1 > W_m = 1/\Delta q_1/\Delta Y_1 > 1/(q/Y)_1$$

or

$$10^{-3} > W_m > 9.4 \cdot 10^{-3} \text{ kg carbon/kg water}$$

The wide interval of a factor of 10 between the limits is disconcerting, but must be accepted in the absence of more elaborate procedures.

Chapter 6

EQUILIBRIUM EFFECTS: LINEAR CHROMATOGRAPHY

6.1 GENERAL FEATURES

Hitherto in our consideration of sorption processes we had assumed uniform initial bed loading and feed conditions. Neither varied with distance or with time. These assumptions are true, or approximately true, in a great many processes and the underlying theory provides important insight into their operation.

Departures from these assumptions occur in a number of ways. Inlet concentrations may vary with time, or initial loading may be nonuniform because of residue from the regeneration step. The regeneration step itself is usually carried out on beds which contain part of the previous mass transfer zone and are therefore nonuniform. Fortunately, as we had seen, desorption processes are insensitive to initial loading, at least under equilibrium conditions, and nonuniformity is therefore not a major disturbance. Fluctuating feed concentrations have a bearing on breakthrough time, but some sense of the effect can be obtained by "bracketing", i.e., by considering the extremes of the fluctuations under steady conditions. In linear systems, an exact treatment is possible, since each point of the disturbance propagates at the same speed fixed by the common Henry constant. This is demonstrated in Illustration 6.2.

In chromatographic operations a narrow band of mixed solutes is deposited at the inlet of the column and subsequently desorbed ("eluted") by passing carrier fluid through the column. The various solutes propagate through the bed in accordance with their retention by the sorbent, the lightly held components moving ahead of the more strongly retained ones. Ultimately this results in a separation of the mixture into pure solute bands or peaks, the so-called chromatogram.

Solutes with nonlinear isotherms interact with each other and give rise to complex behavior. Elaborate theories are available for this case which go beyond the scope of this book. Fortunately, most practical chromatographic processes operate in a region of isotherm linearity for which certain simple statements can be made under equilibrium conditions. They rely on the fact that linear solutes do not interact with each other and furthermore that the concentration fronts propagate *unchanged in shape* through the column: a rectangular pulse initially at the column inlet will maintain its exact shape as it moves through the column, a Gaussian pulse will remain Gaussian of the same width and height, etc. When several linear solutes are deposited near the inlet, they will behave in a similar fashion, each moving independently of the others, and retaining its original pulse shape (Figure 6.1).

This behavior brings about certain simplifications in the treatment of linear systems. We are able to apply, with one or two provisos, the same basic equations which were used under uniform feed and loading conditions, principally the familiar "design" equation used in equilibrium theory:

$$v = z/t = \frac{G_b}{\rho_b \cdot H} = \frac{\rho_f}{\rho_b} \frac{v}{H} \tag{6.1}$$

The proviso is that distance z is measured from the position of the deposited pulse, or a particular concentration within that pulse, rather than the column inlet. Suppose, for example, that we wish to follow the propagation of a rectangular pulse of width Δz located at the inlet. Then the breakthrough time of the leading edge will take the slightly modified form

$$t = \frac{\rho_b}{\rho_f} \frac{H}{v} \cdot (L - \Delta z) \tag{6.2}$$

Δz being suitably adjusted for other positions within the pulse. When several solutes are involved, each one propagates according to its own Henry constant, ultimately leading to a separation into individual pulses or peaks.

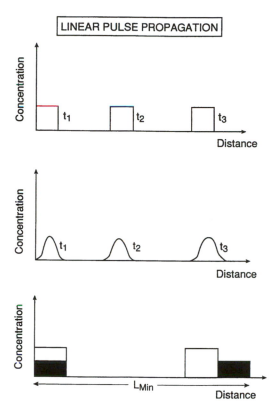

FIGURE 6.1
Propagation of injected solutes with linear isotherms. The bottom graph represents the ideal chromatographic separation of a binary pulse into its components.

We illustrate the application of these principles with the following examples (see Figure 6.1).

Illustration 6.1: Propagation of a Rectangular Pulse

We consider the case of a pulse of width 0.01 m deposited at the inlet of a 1 m column and set $v = 0.1$ m/s, $H = 10^{-2}$. The density ratio is taken to be $\rho_b/\rho_f = 10^4$ because of the use of helium as eluant.

Then for the leading edge, breakthrough occurs at

$$t = \frac{\rho_b}{\rho_f} \cdot \frac{H}{v}(1 - \Delta z) = 10^4 \frac{10^{-2}}{0.1}(1 - 0.01)$$

$$t = 990 \text{ s}$$

The trailing edge lags behind slightly and arrives in 1000 s.

Illustration 6.2: Propagation of Feed Fluctations

A column is being fed with a fluctuating concentration t = f(C) of solute with Henry's constant H. Then the breakthrough time of the leading concentration in a column of length L is given by the usual expression

$$t_{ads} = \frac{\rho_b}{\rho_f} \frac{H}{v} L$$

and that of the concentration level C_1 entering at t_1 by

$$t_{ads} = \frac{\rho_b}{\rho_f} \frac{H}{v} L + t_1$$

In general, a concentration level C will break through at

$$t_{ads} = \frac{\rho_b}{\rho_f} \frac{H}{v} L + f(C)$$

Illustration 6.3: Minimum Column Length Requirement for Separation of a Binary Mixture (Figure 6.1)

Suppose we are attempting to separate two solutes whose Henry constants differ by only 1% ($H_1 = 1$, $H_2 = 1.01$). The minimum column length L required to affect a separation is the one which positions the leading edge of the "light" component at L, and that of the "heavy" component (2) at $L - \Delta z$, where Δz = original band width, set at 0.01 m. We then have:

$$L = \frac{\rho_f}{\rho_b} \frac{v}{H_1} t$$

$$L - \Delta z = \frac{\rho_f}{\rho_b} \frac{v}{H_2} t$$

or

$$\frac{L}{L - \Delta z} = \frac{H_2}{H_1}$$

yielding

$$L = \frac{\Delta z \dfrac{H_2}{H_1}}{\dfrac{H_2}{H_1} - 1} \tag{6.3}$$

Hence

$$L = \frac{0.01 \cdot 1.01}{1.01 - 1} = 1.01 \text{ m}$$

Thus, a little over a meter is required to separate the two solutes. The result is *independent of the shape* of the injected pulse, fluid velocity, or phase densities and depends only upon band width Δz and the ratio of Henry constants. A reduction of pulse width to 1 mm would drop the value of L to 10 cm, while lowering the separation factor to $H_2/H_1 = 1.001$ would raise the column requirement to $L = 10$ m. Evidently these results assume an absence of transport resistance.

Chapter 7

ADSORPTION EQUILIBRIA

In treating this important topic we decided to refrain from the customary recitation of isotherm equations, which are well covered in standard monographs but are not of immediate use in addressing problems confronted by the novice or experienced practitioner. We rejected as well the idea of compiling representative equilibrium isotherm data for reasons of space and their availability, albeit in limited form, in standard texts. We were guided by the fact that in most practical situations the systems under scrutiny have either not been studied before, or are reported in a form unsuitable for immediate application. The demands which arise most frequently are those which require quick approximate assessments of new systems or remedial action in existing systems for which no data are available.

We have attempted to fill this gap by addressing two limiting forms of equilibrium behavior rather than cover the entire operational range of an isotherm. At the lower limit we scrutinize Henry constants. These are frequently needed (and as often unavailable) in analyzing or improving low concentration purification processes. At the other extreme, we have attempted to compile saturation capacities of various sorbents to provide the reader with an upper limit of the capabilities of sorption processes. We are, in a sense, providing a "bracket". Much can be done with these two limiting values and we invite the indulgence of the reader who is unable, with these two parameters at hand, to pinpoint answers that lie in the vast domain between them.

7.1 HENRY CONSTANTS

This notoriously fickle quantity depends not only on temperature and substrate, but also on the detailed nature of the sorbent surface, and the method of measurement or derivation.

Differences of an order of magnitude obtained on the same generic sorbent (e.g., activated carbon) are not uncommon.

We have drawn on three sources of data:

- The precise measurements of low concentration gaseous equilibria on a well-defined sorbent (5 Å zeolite crystals) reported in the text by Ruthven.[2] These data show good internal consistency.
- Henry's constants obtained by fitting a wide range of gaseous isotherm data on activated carbon to reliable isotherm equations, as reported in the text by Valenzuela and Myers.[4] The variability here is considerable.
- Henry's constants H for soil-water systems derived semiempirically from water-octanol partition coefficients K_{OW}, taken from Mackay et al.[10] These values are of relevance in aqueous pollution studies and are not generally available in the conventional adsorption literature. An intermediate value K_{OC}, the partition coefficient for sorption from water on organic carbon contained in the soil can be used to estimate Henry's constants on activated carbon (Table 7.1).

The data are reported in two different ways:

- Henry's constants on zeolite crystals and activated carbon are presented in a plot against normal boiling point, Figure 7.1. There is a tenuous and very coarse correlation between the two if one assumes Henry's constant to be proportional to latent heat of vaporization, which in turn varies directly with boiling point (Trouton's rule). The reader may wish to exploit this. Principally, however, the plot is meant to convey a terse condensation of the available data. We note that the Henry constants are here reported in units of H' [mmol/g · Pa]. This differs from H [kg carrier/kg sorbent] used in preceding chapters for convenience in treating bed dynamics. The conversion is effected through the relation

$$H = H' \times (\text{molar mass of carrier}) \times P_{Tot} \text{ (kPa)}$$

- Henry constants for soil-water systems are given in tabular form (Table 7.1), alongside the octanol-water partition coefficient K_{OW} from which they are derived. Octanol effectively mimics sorption and dissolution in organic matter

TABLE 7.1
Henry Constants for Soil-Water Systems

Solute	log K_{OW}	K_{OW} (m³/m³)	K_{OC} (m³/m³)	$H\left(\dfrac{kg\ water}{kg\ soil}\right)$
Butadiene	1.99	98	39	0.39
n-Pentane	3.45	2800	1100	11
n-Hexane	4.11	13000	5200	52
Cyclohexane	3.44	2800	1100	11
Benzene	2.13	135	54	0.54
Toluene	2.69	490	200	2.0
Styrene	2.88	760	300	3.0
Range of alkanes	2.8–6.25	$630–1.8 \cdot 10^6$	$250–7.1 \cdot 10^5$	2.5–7100
Range of aromatics	2.13–5.52	$140–3.3 \cdot 10^5$	$54–1.3 \cdot 10^5$	5.4–1300
Methyl chloride	0.91	8.1	3.3	0.033
Methylene chloride	1.25	18	7.1	0.071
Chloroform	1.97	93	37	0.37
Carbon tetrachloride	2.64	440	180	1.8
Chlorobenzene	2.80	630	250	2.5
Range of chloroalkanes	0.91–3.93	8.1–8500	3.3–3400	0.033–34
Range of chlorobenzenes	2.80–5.50	$630–3.2 \cdot 10^5$	$250–1.3 \cdot 10^5$	2.5–1300
Range of PCBs	3.90–8.26	$7.9 \cdot 10^3–1.8 \cdot 10^8$	$3.2 \cdot 10^3–7.3 \cdot 10^7$	$32–7.3 \cdot 10^5$

Data from Reference 10.

and is thus a valuable correlating substance used widely in environmental work. The intermediate values K_{OC} denote partition between the aqueous phase and organic carbon OC contained in the soil, and, as previously noted, can be exploited to estimate H for commercial carbons. It is on this carbon phase that adsorption takes place almost exclusively. Conversion among the coefficients, as suggested by Mackay (personal communication), is as follows

$$K_{OC} = 0.4\ K_{OW} \qquad H = 0.01\ K_{OC}$$

Because of the low carbon content of soils (1–5%), Henry constants for soils are low for substances of high and intermediate volatility, ranging from 0.01 – 100 kg water/kg soil. Much higher

HENRY CONSTANTS ON SORBENTS

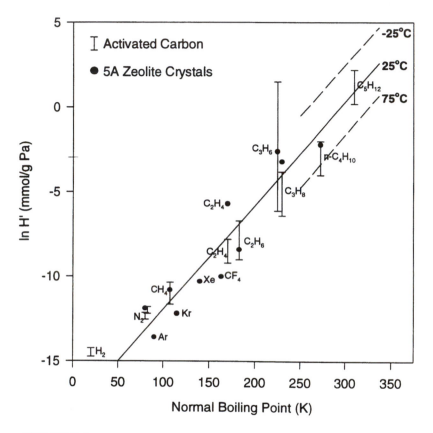

FIGURE 7.1
Henry constants of pure gases on activated carbon and 5 Å zeolite. The abscissa represents the normal boiling point of the gas, which yields a very crude correlating parameter for the Henry constant.

values are seen with high-molecular-weight substances, which tend to be retained tenaciously. A notorious case is that of the PCBs (polychlorinated biphenyls), which may have Henry constants ranging up to H ~10^6 kg water/kg soil, depending on the degree of chlorination.

It should be emphasized that considerable variability attaches to all the coefficients involved, both H and K_{OC}, because of the inevitable variations in the properties of soils and the carbonaceous component contained in them, and to a lesser degree in the correlating octanol-water partition coefficient K_{OW}. The single values reported in Table 7.1 are based on "best average values" compiled by Mackay et al. after an extensive survey of the available data.[10] Variations of an order of magnitude in the values reported by various workers on similar soil samples (and hence similar carbonaceous content) are not uncommon. They parallel the variations in Henry's constants on activated carbon derived from isotherm data and shown in Figure 7.1. Thus, the data of Table 7.1 are to be considered as representative rather than precise values. Much more confidence can be placed in the relative sequence of K_{OC} and H, which generally correctly reproduce the relative retentivity on carbon and soil of the various substrates listed.

7.2   SORPTION SATURATION CAPACITIES

For the purpose of deriving saturation limits we largely drew on the theory of volume filling (potential theory), which shows coincidence of the characteristic curves for various substrates and convergence to a single limiting capacity. The limit is given by 0.48 and 0.43 cc liquid/g adsorbent for activated carbon and silica-gel, respectively.[11,12] This quantity is conveniently broken up into N (mmol/g adsorbent)/ρ_M (mmol/g liquid) from which a linear relation between N and the correlating parameter ρ_M is obtained, Figure 7.2. Liquid densities are computed at a pressure equal to the adsorption pressure but the effect is sufficiently small to allow normal densities to be used throughout.

For the zeolite sorbent, standard saturation capacities reported by the manufacturer were used.

The excellent correlation obtained for both activated carbon and silica gel is due solely to the fact that single values were chosen for the limiting liquid capacities. There is some variability in these values, which we chose to omit for the sake of clarity, but the numbers presented are representative of current good grades of the two sorbents in the equation.

The zeolite data, in contrast, show considerable scatter, having been drawn from manufacturer's data which make no mention of the method of determination and are taken to be approximate values only. The approximation which emerges from the

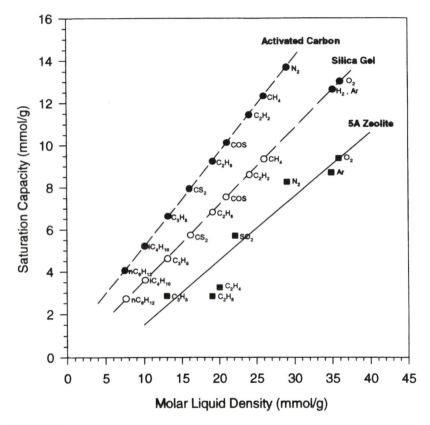

FIGURE 7.2
Saturation capacities of gaseous solutes on three commercial sorbents as a function of solute liquid densities. Based on limiting uptake predicted by potential theory (carbon, silica gel) and manufacturer's data (5 Å zeolite).

presentation is that the limiting capacities stand in roughly the following ratio: C:SiO$_2$:5 Å = 2:1.5:1.

Illustration 7.1: Adsorption of Trace Pollutants from Water on a Carbon Bed. Estimation of Henry's Constants

We consider an aqueous solution carrying traces of n-hexane, benzene, and toluene, which is to be treated by passage through a bed of activated carbon. Henry's constants are to be estimated from K_{OC} values established in environmental studies. The conversion to Henry's constants is given by

$$H\left(\frac{\text{kg water}}{\text{kg carbon}}\right) = K_{OC}\left(\frac{\text{m}^3 \text{ water}}{\text{m}^3 \text{ carbon}}\right)\frac{1}{0.5}\left(\frac{\text{m}^3 \text{ carbon}}{\text{kg carbon}}\right)$$

Using Table 7.1 we obtain:

	$K_{OC}\left(\dfrac{\text{m}^3}{\text{m}^3}\right)$	$H\left(\dfrac{\text{kg water}}{\text{kg carbon}}\right)$
n-Hexane	5,200	10,400
Benzene	54	108
Toluene	200	400

Minimum bed requirement — This will be determined by the substance with the lowest Henry constant, hence

$$W_m = 1/H = 1/108 = 9.3 \cdot 10^{-3} \frac{\text{kg carbon}}{\text{kg water}}$$

Chapter 8

Sorption with Transport Resistance

8.1 GENERAL FEATURES

We turn here to the task of including the effects of mass transfer resistance on the dynamics of the sorptive medium. We do this primarily through the use of what we call design charts, which were derived from similar charts established by Hiester and Vermeulen based on the operative differential equations and which appear in Figures 8.2 to 8.9. In these charts the system behavior is expressed in terms of a number of dimensionless parameters related to fractional saturation and depletion of the medium, its dimensions and transport resistance, time scale and fluid throughput, and the equilibrium isotherm itself. Use of the charts provides a wide range of information related to the response of an initially uniformly loaded or clean sorbent bed to step changes in fluid composition:

- Breakthrough and desorption times
- Concentration profiles and histories
- Length of the mass transfer zone
- Transport coefficients extracted from measured system dynamics
- Minimum bed lengths required to prevent instantaneous breakthrough

The charts have a number of additional attractive features:

- They provide a means of quickly establishing the effect of parameter changes on system performance. A change in transport coefficient, for example, can be immediately

95

translated into adjusted breakthrough times or curves, or bed/purge requirements.

- We can apply the charts to the effective equilibrium curves applicable to inflecting isotherms or adiabatic operations addressed in our previous equilibrium analysis. Unfavorable type III isotherms can also be handled to some extent.

- The dimensionless parameters which appear in the charts can be grafted onto the results of the equilibrium analysis as simple correction factors to account for nonequilibrium effects.

We make mention here as well of the conventional method of evaluating sorption performance, which utilizes a classical NTU-HTU approach, suitably modified to account for the presence of a stationary phase. The results are expressed in terms of the length of the mass transfer zone LMTZ, which equals twice the so-called length of unused bed, LUB. Thus,

$$LMTZ = 2LUB = HTU \cdot NTU = \frac{V}{ka} \int_1^2 \frac{dC}{C^* - C} \qquad (8.1)$$

where V is the propagation velocity of the front. C and ka refer to fluid or solid phase concentrations and transfer coefficients, depending on which mode of transfer is chosen. For solid phase conditions prevailing for example, and using expressions given in the preceding equilibrium analysis for propagation velocity

$$V = \frac{G_b}{\rho_b \Delta q / \Delta Y} = \frac{v_f \rho_f}{\rho_b \cdot \Delta q / \Delta Y} \qquad (8.2)$$

we obtain the expressions

$$LMTZ = \underbrace{\frac{V}{k_s a}}_{HTU} \underbrace{\int \frac{dq}{q^* - q}}_{NTU} = \underbrace{\frac{G_b (= v_f \rho_f)}{\rho_b \Delta q / \Delta Y \, k_s a}}_{HTU} \underbrace{\int_{q_1}^{q_2} \frac{dq}{q^* - q}}_{NTU} \qquad (8.3)$$

The NTU integral is evaluated in the usual fashion by using driving forces $q^* - q$ given by the vertical distance between equilibrium and operating line. Both NTU and HTU refer to solid phase conditions, where $k_s a$ (s^{-1}) = solid phase mass transfer coefficient. Similar expressions, involving $k_p a$ and Y^*-Y, apply when fluid conditions in the particle pores prevail. In this case the driving force is evaluated from the *horizontal* distance between equilibrium and operating lines.

The above expressions are used mainly for design purposes, i.e., to estimate the length of bed required for a particular adsorption operation with a prescribed on stream time of t. The pertinent expression is obtained by adding half of the transfer zone length to the length of bed saturated under equilibrium conditions:

$$\text{Total length of bed } L = \frac{1}{2}\text{LMTZ} + \frac{G_b\left(= v_f \rho_f\right)}{t \cdot \rho_b \, \Delta q / \Delta Y} \qquad (8.4)$$

It is evident from the foregoing that much less information is to be gained by applying the NTU-HTU concept than can be gleaned from the design charts. Its primary use is in design or in the estimation of transport parameters $k_s a$ ($k_f a$) from measured LMTZ values. It fails to establish, among other things, the conditions required to prevent instantaneous breakthrough or to improve performance. More importantly, it is restricted to favorable type I isotherms. Neither linear isotherms (an important category) nor unfavorable ones or those with inflections are accessible by the HTU-NTU approach. Even in applying it to type I isotherms, the assumption has to be made that a constant shape ("constant pattern") has been attained, i.e., no account is taken of the early stages of profile development. This can only be done by means of the design charts which cover the entire range of operational conditions. Nevertheless, the HTU-NTU approach retains its popularity mainly on the strength of its resemblance to well-known and established practices. We hope by the introduction of our design charts to open up wider fields of application and to allow a more profound and at the same time more rapid analysis to be made.

We may, however, be disappointed.

8.2 THE SEPARATION FACTOR r

Use of the design charts requires us to express equilibrium
isotherms in terms of a parameter which can be introduced into
the charts. This is done by means of the separation factor r which
is the reciprocal of the relative volatility known from distillation
theory.

$$r = \frac{(\Delta Y/\Delta Y_o)(1 - \Delta q/\Delta q_o)}{(\Delta q/\Delta q_o)(1 - \Delta Y/\Delta Y_o)} \tag{8.5}$$

Here ΔY and Δq denote the interval from the start of the
equilibrium curve to a particular point on it. ΔY_0 and Δq_0
encompass the full range of the curve. This is demonstrated by
the numerical values displayed in the accompanying graph,
Figure 8.1. Values of $r < 1$ ($\alpha > 1$) represent type I isotherms
for adsorption, type III isotherms for desorption, i.e., situations
which give rise to sharp sorption fronts. Values of $r > 1$, on the
other hand, lead to broadening fronts, i.e., those arising from
desorption along type I, and adsorption along type III isotherms. $r = 1$ represents the threshold value for a linear isotherm, while $r = 0$ leads to the limiting case known as an
irreversible isotherm. Sharply rising isotherms of the type
shown by H_2O on zeolitic sorbents or high-boiling solvent
vapors on activated carbon effectively have separation factors
which approach this limit. Intermediate favorable isotherms,
such as those displayed by low-boiling hydrocarbons on carbon, fall in the range $r = 0.7$–0.3, with $r = 0.5$ being a good
initial guess. $r = 1$ is invoked at low concentrations, i.e., whenever Henry's law applies. More precise values are best obtained
by matching the isotherm segment spanned by the operating
line to separation factors shown. Given the uncertainties in
reported equilibrium isotherms, as well as deviations from the
required symmetry, one should not put a high premium on
obtaining values of great precision. Most practical processes
operate in regions of low sensitivity to r values, thus a precision
of ± 0.05 is quite acceptable.

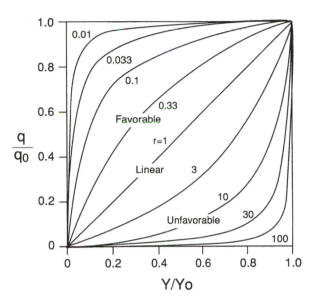

FIGURE 8.1

Non-dimensionalized equilibrium isotherms ploted in terms of the separa-

tion factor $r = \dfrac{\Delta Y/\Delta Y_o \left(1 - \Delta q/\Delta q_o\right)}{\Delta q/\Delta q_o \left(1 - \Delta Y/\Delta Y_o\right)}$. Langmuir (type I) isotherms have r < 1,

unfavorable (type III) r > 1. In the linear (Henry's law) region, r = 1.

8.3   TRANSPORT COEFFICIENTS

This is a notoriously difficult subject, given that both film and particle resistances may be operative. The latter in particular is often a complex composite of various diffusional modes, e.g., surface and pore diffusion, Knudsen diffusion, hindered diffusion into zeolitic cavities which depend on concentration and temperature. Liquid sorption transport has its own idiosyncracies. In addition, both fluid and solid phase driving forces may be operative. To unravel these complexities, we make the following observations based on extensive surveys of both gas and liquid phase sorption processes.

Under practical conditions, film resistance rarely plays a role in vapor phase sorption, and does so only infrequently or to a minor extent in liquid phase applications. Particle diffusion represents, for the most part, the major resistance in practical sorption processes. Its value fluctuates but in our survey, particularly of conventional purification processes (drying of air, removal of low boiling impurities), we found a convergence to the same order of magnitude range. More severe aberrations occur in certain instances but for the most part we feel encouraged in providing the following guidelines which are conveniently summarized in the table below. Both gas and liquid phase applications are considered, as well as values based on fluid and solid phase driving forces. Particle radius is set at 1 mm. To adjust it to other radii, use is made of the Glueckauf relations which transform diffusivities into mass transfer coefficients:

$$k_f a = 15 \ D_f / R^2$$

$$k_s a = 15 \ D_s / R^2$$

$$(8.6)$$

where the subscripts denote fluid and solid phase transport, respectively.

The suggested values, to be used for tentative estimates, are as follows:

TABLE 8.1
Range of Transport Coefficients

	Gas Phase		Liquid Phase	
Diffusivity (cm²s⁻¹)	$D_p \cong 10^{-2}\text{–}10^{-3}$	$D_s \cong 10^{-5}\text{–}10^{-6}$	$D_p \cong 10^{-5}\text{–}10^{-6}$	$D_s \cong 10^{-7}\text{–}10^{-8}$
Mass Transfer Coefficient (s⁻¹)	$k_p a \cong 10^{-1}$	$k_s a \cong 10^{-2}\text{–}10^{-3}$	$k_p a \cong 10^{-2}\text{–}10^{-3}$	$k_s a \cong 10^{-4}\text{–}10^{-5}$

Deviations may occur due to unusually high contributions by surface diffusion, an unusual lowering of diffusivity by hindered transport, e.g., into a zeolite, or other reasons. Some unusual deviations of this type occur in the drying of organic solvents with zeolites (see Illustration 8.7). In the end one must draw on direct diffusion experiments or diffusivities extracted from breakthrough curves for more reliable values.

8.4   THE DESIGN CHARTS

These are presented in terms of the following dimensionless parameters:

TABLE 8.2
Chart Parameters

- *The separation factor r,* representative of the isothermal or effective equilibrium curve

- Dimensionless distance N, equivalent to number of transfer units:

$$N = k_p a \frac{z}{v} f(r) \quad \text{or} \quad k_s a \frac{z}{v} \frac{\rho_b}{\rho_f} \frac{\Delta q}{\Delta Y} f(r)$$

- Dimensionless Time T:

$$T = k_s a \, tf(r) \quad \text{or} \quad k_p a \, t \frac{\rho_f}{\rho_b} \frac{\Delta Y}{\Delta q} f(r)$$

- Fractional fluid phase concentration $\Delta Y / \Delta Y_o$
- Fractional solid phase concentration $\Delta q / \Delta q_o$

Here ρ_b and ρ_f represent bed and fluid densities, respectively, their ratio being approximately 10^3 and 1 for gases and liquids at standard conditions. $\Delta q / \Delta Y$ is the slope of the operating line connecting initial and feed (or plateau) points, $k_p a$ and $k_s a$ are the mass transfer coefficients related to diffusivities via the Glueckauf relations (8.6), ΔY_o and Δq_o the full saturation values over the operative range of the equilibrium curve. f(r) is a correction factor for isotherm nonlinearity: $f(r) = 2/1+r$ in the favorable region and $f(r) = r^{-1/2}$ in the unfavorable region.

Five charts are presented for fractional fluid concentration of $\Delta Y / \Delta Y_o = 0.01, 0.1, 0.5, 0.9,$ and 0.99 (Figures 8.2 to 8.6). These are meant primarily for use in adsorption operations and allow, among other things, the rapid derivation of full breakthrough curves. For desorption operations, it is of greater relevance to monitor fractional adsorbed phase concentrations, hence we present additional charts which combine the breakthrough levels $\Delta q / \Delta q_o = 0.01, 0.1, 0.9,$ and 0.99 (Figures 8.7 to 8.9). For desorption, one uses the relation $(1 - \Delta q / \Delta q_o)$. Thus, a breakthrough level of 0.9 corresponds to a residual loading level of $1 - 0.9 = 0.1$ during desorption.

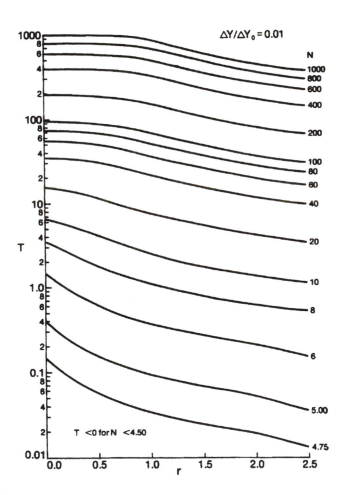

FIGURE 8.2
Chart relating fractional fluid phase concentration $\Delta Y/\Delta Y_0 = 0.01$ to nondimensionalized distance N, time T, and equilibrium uptake r. (Reproduced
from Basmadjian, D., *Advances in Drying*, Vol. 3, p. 333. Taylor & Francis,
Washington, D.C., 1984. Reproduced with permission. All rights reserved.)

FLUID PHASE CHART

10% BREAKTHROUGH

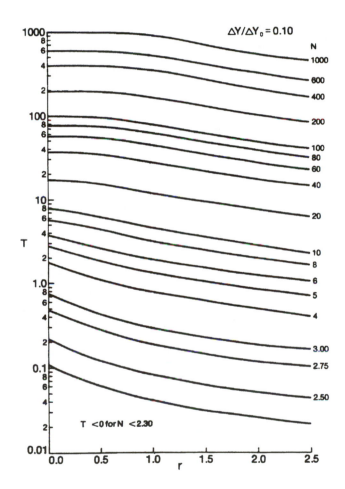

FIGURE 8.3
Chart relating fractional fluid phase concentration $\Delta Y/\Delta Y_o = 0.1$ to nondimensionalized distance N, time T, and equilibrium uptake r. (Reproduced from Basmadjian, D., *Advances in Drying*, Vol. 3, p. 334. Taylor & Francis, Washington, D.C., 1984. Reproduced with permission. All rights reserved.)

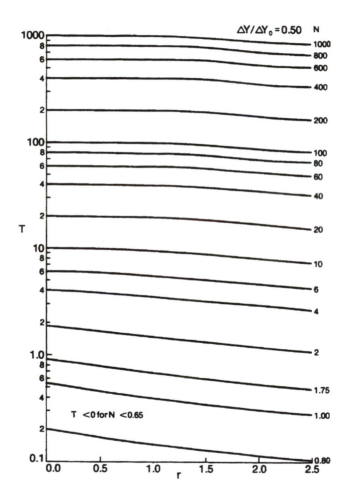

FLUID PHASE CHART

50% BREAKTHROUGH

FIGURE 8.4
Chart relating fractional fluid phase concentration $\Delta Y/\Delta Y_o = 0.5$ to nondi-mensionalized distance N, time T, and equilibrium uptake r. (Reproduced from Basmadjian, D., *Advances in Drying*, Vol. 3, p. 335. Taylor & Francis, Washington, D.C., 1984. Reproduced with permission. All rights reserved.)

FLUID PHASE CHART

90% BREAKTHROUGH

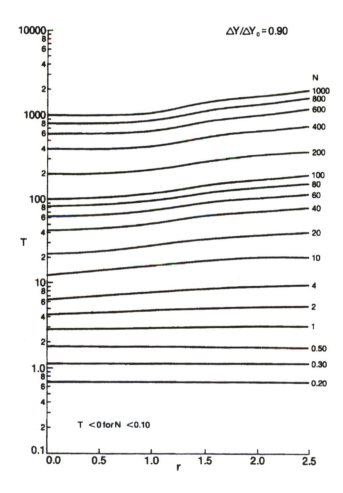

FIGURE 8.5
Chart relating fluid phase concentration $\Delta Y/\Delta Y_0 = 0.9$ to nondimensionalized distance N, time T, and equilibrium uptake r. (Reproduced from Basmadjian, D., *Advances in Drying,* Vol. 3, p. 336. Taylor & Francis, Washington, D.C., 1984. Reproduced with permission. All rights reserved.)

FLUID PHASE CHART

99% BREAKTHROUGH

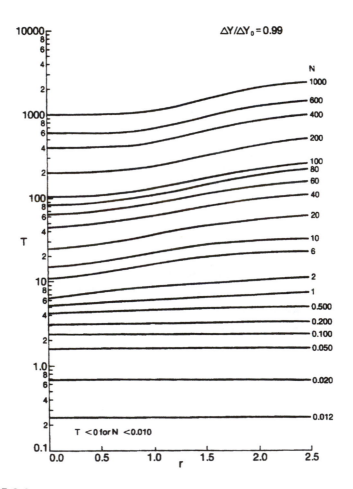

FIGURE 8.6
Chart relating fluid phase concentration $\Delta Y/\Delta Y_o = 0.99$ to nondimension-alized distance N, time T, and equilibrium uptake r. (Reproduced from Basmadjian, D., *Advances in Drying,* Vol. 3, p. 337. Taylor & Francis, Washington, D.C., 1984. Reproduced with permission. All rights reserved.)

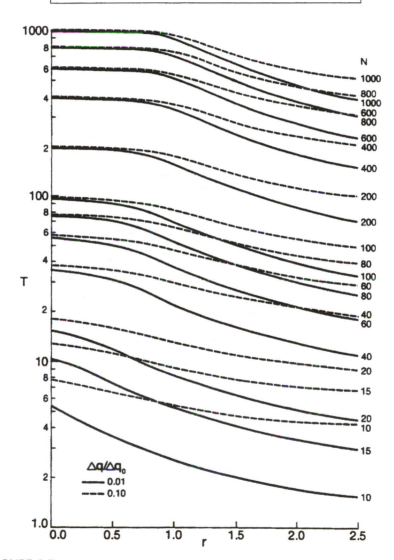

IGURE 8.7

Chart relating fractional fluid phase concentration $\Delta q/\Delta q_o = 0.01$ and 0.1 to nondimensionalized distance N, time T, and equilibrium uptake r. (Reproduced from Basmadjian, D., *Advances in Drying*, Vol. 3, p. 338. Taylor & Francis, Washington, D.C., 1984. Reproduced with permission. All rights reserved.)

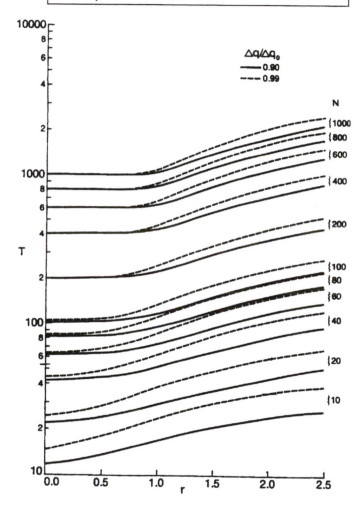

FIGURE 8.8
Chart relating fractional fluid phase concentration $\Delta q/\Delta q_o = 0.9$ and 0.99 to nondimensionalized distance N, time t, and equilibrium uptake r. (Reproduced from Basmadjian, D., *Advances in Drying*, Vol. 3, p. 339. Taylor & Francis, Washington, D.C., 1984. Reproduced with permission. All rights reserved.)

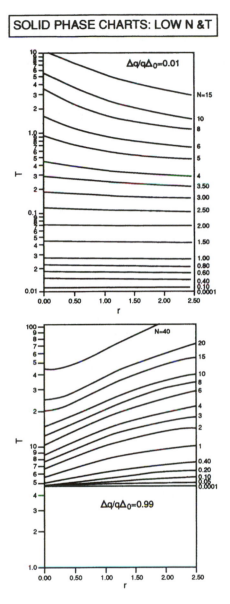

FIGURE 8.9
Chart relating fluid phase concentration $\Delta q/\Delta q_o$ = 0.01 and 0.99 at low values of nondimensionalized distance N, time t, and equilibrium uptake r.

8.5   CONDITIONS FOR PREVENTING INSTANTANEOUS BREAKTHROUGH

These are determined by the relations inserted at the bottom left side of the fluid phase charts. To illustrate their use, we note that at the $\Delta Y/\Delta Y_o = 0.01$ level, dimensionless distance must be $N \geq 4.5$. Thus, to prevent instantaneous breakthrough at the start of an adsorption operation we must have one of the following conditions fulfilled at the 0.01 level:

$$\text{Bed height } z > 4.5 \cdot \frac{v}{k_p a}$$

$$\text{Velocity } v < 0.22/z \cdot k_p a$$

(8.7)

Similar relations apply for solid phase transport.

The means we have at our disposal, therefore, for the prevention of this phenomenon is one or both of the following: increase in bed height or decrease in fluid velocity. These conditions agree with physical reasoning, but are placed on a quantitative basis by the relations given above.

8.6   GRAFTING OF TRANSPORT EFFECTS ONTO EQUILIBRIUM THEORY (NO TRANSPORT RESISTANCE)

This is achieved by multiplying the operational parameters calculated by equilibrium theory (breakthrough and desorption times, minimum bed and purge requirements) by a correction factor composed of the ratio of dimensionless distance N and dimensionless time T. Division of these two chart parameters leads to the expression

$$\frac{N}{T} = \left[\frac{z\rho_b}{\rho_f \cdot v}\right]\frac{\Delta q}{\Delta Y} = \left[\frac{\text{kg bed}}{\text{kg dry gas}}\right]\frac{\Delta q}{\Delta Y}$$

(8.8)

which upon rearrangement yields

$$W_a \left[\frac{kg\ bed}{kg\ dry\ gas} \right] = \frac{N}{T} \cdot \left[\frac{\Delta q}{\Delta Y} \right]^{-1} = \frac{N}{T} W_m \qquad (8.9)$$

Thus, the actual bed requirement W_a is seen to equal the minimum bed requirement W_m obtained by equilibrium theory, multiplied by the correction factor N/T. Similar corrections can be applied to other operational parameters, which we summarize for convenience below.

TABLE 8.3
Bed Parameters With and Without Transport Resistance

Parameter	Equilibrium Theory	Transport Resistance Added
Breakthrough time (type I isotherm)	$t_e = \dfrac{\rho_b\ \Delta q/\Delta Y}{\rho_f v} L$	$t_a = \left(\dfrac{T}{N} \right) \cdot t_e$
Desorption time (type III isotherm)	$t_e = \dfrac{\rho_b\ \Delta q/\Delta Y}{\rho_f v} L$	$t_a = \left(\dfrac{T}{N} \right) \cdot t_e$
Bed requirement (type I isotherm)	$W_m = (\Delta q/\Delta Y)^{-1}$	$W_a = \left(\dfrac{N}{T} \right) W_m$
Purge requirement (type III isotherm)	$G_m = (\Delta q/\Delta Y)$	$G_a = \left(\dfrac{T}{N} \right) G_m$

Application of these relations will be demonstrated in a number of illustrations to follow.

8.7   CONDITIONS FOR THE ATTAINMENT OF SHARP SORPTION FRONTS

Ideally one would want the sorption front to occupy as little space in the sorbent bed as possible, i.e., LMTZ to be near or at zero (discontinuous front). The design charts provide a means for estimating parameter values which lead to this desirable condition. Briefly, it requires equality to be attained between dimensionless time and distance:

$$N = T \qquad (8.10)$$

For broadening fronts with r > 1, the charts show that this condition cannot be met except in the limit r = 1. This is in agreement with physical reasoning. In the range $0 \leq r \leq 1$, on the other hand, near-equality is attained at N = 100 for $0 \leq r \leq 0.5$, and full equality over the entire range at approximately N = 1000. This means that the ratio of bed height to fluid velocity must, to satisfy the condition of a sharp front, attain the range of values:

$$\frac{z}{v} = \frac{10^2}{k_p a} \text{ to } \frac{10^3}{k_p a} \tag{8.11}$$

For gas phase applications, this may be realized by setting $\frac{z}{v} = 10 - 10^3$ s, of which the lower limit in particular is within reasonable reach of practical realization (e.g., z = 5 m, v = 0.4 m/s). Liquid systems require much higher ratios: $\frac{z}{v} = 10^4 - 10^6$ s, but these are not beyond realization, given that liquid velocities in the range $10^{-3} - 10^{-4}$ m/s are often employed. Thus, bed heights of 1 – 10 m would satisfy the lower limit of the z/v range. Evidently operating conditions may arise which must accommodate higher fluid velocities, or risk excessive bed diameters. One must then live with transfer zones of considerable length. The above relations, however, indicate to what extent adjustments in z and v can be made to reduce mass transfer effects.

We now demonstrate the use of the design charts and the various relations given above by presenting a number of practical illustrations. In most cases we have deliberately omitted experimental data, although these may be available in the open literature. The problems are instead posed in a raw form, as is almost invariably the case during in-house or external consultations.

Illustration 8.1: Removal of Methane from Air

One hundred parts per million by weight CH_4 in air is to be removed by adsorption on a zeolite. A bed height of 1 m and superficial velocity of 10 m/s is proposed. The request is for a preliminary analysis of the process.

We limit ourselves to an investigation of possible temperature effects and an estimate of bed heights and velocities needed to prevent instantaneous breakthrough. Recommendations are then made which would lead to a more comprehensive study.

Temperature rise ΔT — From Equation (4.2)

$$\Delta T = Y_F \, \Delta H / C_{pb}$$

Using the suggested typical values of $\Delta H \cong 2500$ kJ/kg, $C_{pb} \cong 1$ kJ/kg K we obtain

$$\Delta T = 10^{-4} \cdot 2500/1 = 0.25\,°C$$

Isothermal operation may therefore be assumed.

Mass transfer coefficient — Since we are dealing with a low-molecular-weight sorbate with the dominant transport events likely confined to the pore space, we choose from Table 8.3

$$k_p a = 10 \ s^{-1}$$

Instantaneous breakthrough — At the 0.01 fractional breakthrough level, the condition to be satisfied to prevent premature breakthrough is

$$N = k_p a \frac{z}{v} \geq 4.5$$

Solving for z or v this translates into the conditions

$$z > 4.5 \ m \qquad v < 4.5 \ m/s$$

Suggested remedies — (1) Raise bed height to 8–10 m. This value may require excessive adsorbent inventory and/or pressure drop; (2) reduce velocity by compressing the feed to 10 atm, with an attendant velocity of 1 m/s. Compression costs become part of the overall operating costs.

Verdict — Both alternatives appear worth pursuing. In addition, activated carbon should be considered because of its higher capacity (factor of 2 at saturation).

Illustration 8.2: Removal of Butane from N_2–H_2

A low concentration of butane is to be removed by adsorption on activated carbon from a stream of N_2–H_2 (25/75%). Molar mass of carrier = 8.75, so that ρ_b/ρ_f = 450/0.3 = 1500. Feed concentration Y_F = 10^{-3} kg/kg with a corresponding loading of $2 \cdot 10^{-2}$ kg/kg carbon. Bed height and velocity are set at z = 2 m and v = 1 m/s. A full analysis is to be carried out.

Henry constant — From Figure 7.1, (ln H')avg = –3; H' = 0.050 mmol/g · Pa:

$$H = H' \text{ (molar mass of carrier) } P_{tot} \text{ (kPa)}$$

$$H = 0.050 \cdot 8.75 \cdot 100 = 43.8 \text{ kg air/kg carbon}$$

Operating line —

$$\frac{\Delta q}{\Delta Y} = \frac{q_F - 0}{Y_F - 0} = \frac{2 \cdot 10^{-2}}{10^{-3}} = 20$$

Thus, the operating line slope is smaller than the Henry constant, as required.

Temperature rise ΔT —

$$\Delta T = Y_F \, \Delta H/Cp = 10^{-3} \cdot 2500/1 = 2.5\,°C$$

Operation may still be considered isothermal.

Mass transfer coefficient — We retain $k_p a$ = 10 chosen for CH_4, since C_4H_{10}, may still be considered a light gas. Molecular weight adjustment was omitted because its effect probably is counterbalanced by surface diffusion which may be expected to contribute to the overall diffusivity.

Instantaneous breakthrough — As before, instantaneous breakthrough occurs when $N = k_p a \dfrac{z}{v} \leq 4.5$ at the 0.01 level. To avoid this, we must satisfy the condition

$$z > 0.45 \text{ m} \qquad v < 2.2 \text{ m/s}$$

Both conditions are fulfilled. Note that if the lower $k_p a$ value had been chosen, the conditions would not be met and z and/or v would have to be adjusted accordingly.

Adsorption breakthrough time — In the absence of complete equilibrium data, we assume a separation factor r = 0.5. Other values are considered below for comparison and show a general insensitivity to the value of r.

$$N = k_p a \frac{z}{v} \frac{2}{1+r} = 10 \cdot \frac{2}{1} \cdot \frac{2}{1.5} = 22.6$$

From the fluid phase chart, Figure 8.3, at $\Delta Y / \Delta Y_o = 0.1$, T = 17.5. The breakthrough time at equilibrium conditions is given by

$$t_e = \left[\frac{\rho_b}{\rho_f} \right] \frac{\Delta q / \Delta Y}{v} = 1500 \cdot \frac{20}{1} = 30,000 \text{ s} = 8.3 \text{ h}$$

The actual time is then obtained by multiplying t_e by the correction factor T/N (Table 8.3)

$$t_a = t_e \cdot T/N = 3 \cdot 10^4 \cdot \frac{17.5}{22.6} = 23,230 \text{ s} = 6.5 \text{ h}$$

Conditions obtained at other values of r are summarized below:

r	T at N = 22.6	T/N	t (h)
0	19	0.84	7.0
0.2	18	0.80	6.7
0.7	17	0.75	6.3

One notices that the variations in t_a due to change in isotherm shape are not severe. This is due to the fact that the front is fairly sharp. Thus, the precise value of the separation factor or isotherm curvature is not of great consequence.

Length of mass transfer zone LMTZ — Here the procedures are inverted, T being first calculated at a particular instant, say t = 10,000 s, and corresponding values of N read from the charts, from which bed distances are then extracted. Thus,

$$T = k_p a \cdot t \cdot \frac{\rho_f}{\rho_b} \cdot \frac{\Delta Y}{\Delta q} f(r) = 10 \cdot 10^4 \cdot \frac{1}{1500} \cdot \frac{1}{20} \frac{2}{1+0.5} = 4.4$$

We calculate LMTZ for both the fluid and solid phases between the fractional breakthrough concentrations 0.01 and 0.99. The low value of T signals the fact that we can expect long transfer zones, which turns out to be the case. Recall that for sharp fronts, the minimum requirement is $T = N = 100$.

Fluid phase LMTZ: Using the fluid phase charts, Figures 8.2 and 8.6, and the relation for N given in Table 8.2, we obtain

$$N_{0.01} = 17 \quad z = \frac{N}{k_p a} \frac{v}{f(r)} = \frac{17}{10} \cdot \frac{1}{2/1.5} = 1.28 \text{ m}$$

$$N_{0.99} = 4 \quad z = \frac{N}{k_p a} \frac{v}{f(r)} = \frac{4}{10} \cdot \frac{1}{2/1.5} = 0.33 \text{ m}$$

Thus, at time $t = 10{,}000$ s, the fractional fluid concentration is 0.99 at 0.33 m from the inlet, and 0.01 at 1.28 m from the inlet, yielding a total LMTZ of $1.28 = 0.33 = 0.95$ m.

Solid phase LMTZ: We retain the same value of T, but revert to the solid phase charts for the calculation of N. Because of the low value of T, we use the low range charts, Figure 8.9, and obtain

$$N_{0.01} = 11 \quad z = \frac{N}{k_p a} \frac{v}{f(r)} = \frac{11}{10} \cdot \frac{1}{2/1.5} = 0.825 \text{ m}$$

$$N_{0.99} = 0 \quad z = \frac{N}{k_p a} \frac{v}{f(r)} = \frac{0}{10} \cdot \frac{1}{2/1.5} = 0 \text{ m}$$

Thus, while the LMTZ is approximately the same as in the fluid phase case (0.825 m vs. 0.95 m), the solid phase zone is shifted toward the inlet, 99% solid phase saturation prevailing at the inlet.

Bed Requirement W_a: Bed requirement depends on the allowable fractional breakthrough level to define exhaustion of the bed. We choose $\Delta Y / \Delta Y_o = 0.1$ for which we had calculated $T = 17.5$ and $N = 22.6$. Using the relation given in Table 8.3 we obtain

$$W_a = \frac{N}{T} W_m = \frac{N}{T} \frac{1}{\Delta q/\Delta Y} = \frac{22.6}{17.5} \cdot \frac{1}{20} = 0.065 \text{ kg bed/kg carrier}$$

Desorption time (hot purge) — In the absence of a set of butane isotherms, we are unable to determine accurately the adiabatic desorption curve, and values of its operating line slope $\Delta q/\Delta Y$ and separation factor r. Instead we argue as follows: $\Delta q/\Delta Y$ must be less than the value for adsorption, $(\Delta q/\Delta Y)_{ads} = 20$. This is seen from the adiabatic desorption diagram, Figure 4.5. We choose $(\Delta q/\Delta Y)_{des} = 5$, accompanied by an average separation factor of r = 0.5. By making this choice, we are not only able to make the necessary computations using the design charts, but we can immediately fix the degree of enrichment obtained during desorption, which will be by a factor of 4, identical to the ratio of the desorption/adsorption operating line slopes. The only factor we are unable to compute is the regeneration temperature required to achieve this result. That calculation would require a full complement of butane isotherms over a range of temperatures. Experience with similar substrates indicates, however, that a purge temperature of 200°C should be sufficient to bring about the desired result.

Desorption time — To carry out the necessary calculations, we retain the value of N = 22 (i.e., the same mass transfer coefficient as in adsorption) and a use velocity of 1 m/s. This implies that half the feed or product, after being heated, is diverted to the desorption process.

To signal the end of desorption, we use the chart for $\Delta q/\Delta q_o = 0.9$, Figure 8.5, which corresponds to a fractional undesorbed residue on the solid of 0.1. For N = 22, we obtain a value of T = 24. Hence,

$$T = 24 = k_p a \cdot t \cdot \frac{\rho_f}{\rho_b} \frac{\Delta Y}{\Delta q}$$

and the desorption time will be given by

$$t = T \frac{1}{k_p a} \cdot \frac{\rho_b}{\rho_f} \cdot \frac{\Delta q}{\Delta Y} = 24 \cdot \frac{1}{10} \cdot 1500 \cdot 5 = 18,000 \text{ s}$$

The adsorption step required 23,230 s, so that some 5000 s are available to cool the bed and prepare it for the next adsorption step.

Purge requirement — The appropriate expression is taken from Table 8.3 and yields (at the 0.1 level)

$$G_a = \frac{T}{N} \cdot G_m = \frac{24}{22} \cdot 5 = 5.5 \text{ kg purge/kg bed}$$

Illustration 8.3: Extraction of Mass Transfer Coefficient from Measured Breakthrough Data

In a liquid phase adsorption process using a bed height z of 2 m, and fluid velocity v of 10^{-3} m/s, breakthrough at the 0.01 fractional level occurred after approximately 10^6 s (~280 h). Such long breakthrough times are typical of liquid phase applications.

Isotherm data yielded a separation factor r = 0.5 over the span of the operating line which had a slope of 10^3. We wish to compute the effective mass transfer coefficient $k_s a$ for the system.

A convenient way to proceed is to use the ratio N/T, which eliminates the unknown quantity $k_s a$. Thus,

$$\frac{N}{T} = \frac{z}{vt} \cdot \frac{\rho_b}{\rho_f} \cdot \frac{\Delta q}{\Delta Y} = \frac{2}{10^{-3} \cdot 10^6} \cdot 1 \cdot 10^3 = 2$$

On inspection of the chart for $\Delta Y/\Delta Y_o = 0.01$ (Figure 8.2) this ratio is found to occur at an approximate value of N = 15 (r = 0.5). Hence,

$$N = k_s a \cdot \frac{z}{v} \cdot \frac{\rho_b}{\rho_f} \cdot \frac{\Delta q}{\Delta Y} f(r) = 15$$

and

$$k_s a = \frac{15}{2/10^{-3} \cdot 1 \cdot 10^3} \frac{15}{2} = 5.6 \cdot 10^{-6} \text{s}^{-1}$$

This compares with a range of $10^{-4} - 10^{-5}$ s^{-1} recommended for use in initial estimates of $k_s a$ (Table 8.1).

Illustration 8.4: Adsorption of Trace Pollutants from an Aqueous Solution onto Activated Carbon (Revisited)

The term trace pollutants signals the fact that they may fall in the Henry's law range. Henry constants for organic pollutants typically vary over the range $H = 10^2 - 10^5$ kg water/kg carbon. We consider the case where $H = 10^3$ corresponds to the "lightest" component, i.e., the pollutant which breaks through first. Heavier components follow in a well-ordered sequence of sharp fronts in order of increasing Henry constants. A limited version of this case was considered in Illustration 5.3 under equilibrium conditions.

In designing the bed, we wish to minimize transfer zone length and consequently aim at a value of $N = 100$. We set bed height at $z = 1$ m, velocity at $v = 10^{-3}$ m/s, both typical values for these applications. The density ratio is rounded off to $\rho_b/\rho_f = 1$. For $k_s a$ we choose the higher value of the suggested range, 10^{-4} s^{-1}. We obtain

$$N = k_s a \frac{z}{v} \cdot \frac{\rho_b}{\rho_f} H = 10^{-4} \cdot \frac{1}{10^{-3}} \cdot 1 \cdot 10^3 = 100$$

where the Henry constant H now takes the place of the slope of the operating line $\Delta q/\Delta Y$. The corresponding value for T is read from the chart for $\Delta Y/\Delta Y_o = 0.01$ (Figure 8.2):

$$T = 72 = k_s a \cdot t$$

and

$$t = \frac{72}{10^{-4}} = 7.2 \cdot 10^5 \, s$$

Thus, the bed will be exhausted after 200 h. The bed requirement itself is quickly available from Table 8.3:

$$W_a = \frac{N}{T} \cdot (H)^{-1} = \frac{100}{72} \cdot 10^{-3} = 1.3 \cdot 10^{-3} \text{ kg carbon/kg water}$$

What would happen if the feed concentrations extended into the nonlinear zone? We can venture an estimate: the operating line slope $\Delta q/\Delta Y$ must be less than Henry's constant, e.g., $0.3 \cdot 10^3$. This yields $N = 30$ and the corresponding dimensionless time read from the chart at $r = 0.5$ drops to $T \cong 22$. Hence, breakthrough time occurs about three times earlier.

$$t = \frac{T}{k_s a} = \frac{22}{10^{-4}} = 2.2 \cdot 10^5 \, s$$

and the bed requirement increases correspondingly to

$$W_a = \frac{N}{T}\left(\frac{\Delta q}{\Delta Y}\right)^{-1} = \frac{30}{22}\frac{1}{0.3 \cdot 10^3} = 4.6 \cdot 10^{-3} \text{ kg carbon/kg water}$$

Illustration 8.5: Clearance of a River Bed

A stretch of river bed 1000 m long has been contaminated with a pollutant. Velocity of the river is estimated at 0.1 m/s.
We consider two cases:

(I) $H = 10^3$, $k_s a = 10^{-5}$ s, typical of stubbornly held chlorinated hydrocarbons (see Table 7.1).
(II) $H = 0.1$, $k_s a = 10^{-4}$ s, typical of the more volatile hydrocarbons.

The parameters are such that they call for the use of both the high and low range $\Delta q/\Delta q_o$ charts (Figures 8.7 to 8.9).
$H = 10^3$ kg water/kg sediment, $k_s a = 10^{-5}$ s^{-1} — Dimensionless distance:

$$N = k_s a \cdot \frac{z}{v} \cdot \frac{\rho_b}{\rho_f} \cdot H = 10^{-5} \cdot \frac{10^3}{10^{-1}} \cdot \frac{2.5}{1} \cdot 10^3 = 250$$

The corresponding value of dimensionless time T is read off the high range chart at the level of $\Delta q/\Delta q_o = 0.99$ (Figure 8.8):

$$T = k_s a \, t = 250$$

and

$$t = \frac{250}{10^{-5}} = 2.5 \cdot 10^7 \text{s} = 290 \text{ days}$$

Clearance here is faster than in the case of a groundwater purge (Illustration 2.5: 1010 days) because of the much higher water velocity. The effect, however, is attenuated by the increased length to be cleared and the effect of transport resistance, which was absent in the previous example.

H = 0.1 kg water/kg sediment, $k_s a = 10^{-4} s^{-1}$ — Here the dimensionless distance is diminished by a factor of 10^{-3}, so that N = 0.25. The corresponding value of T is read from the low range chart, Figure 8.9, and yields

$$T = k_s a \; t = 5.2$$

and

$$t = \frac{5.2}{10^{-4}} = 5.2 \cdot 10^4 \text{s} = 0.6 \text{ days}$$

Clearance here is very rapid and normal sediment conditions are quickly restored.

Illustration 8.6: Prediction of Binary Breakthrough Curves

We attempt here to match experimental breakthrough data reported for the adsorption of benzene/hexane vapors in helium onto silica gel (Shen and Smith[13]). Pertinent parameters reported by the authors are listed below:

$$L = 0.02 \text{ m} \qquad v = 0.0265 \text{ m/s} \qquad D_p = 2.8 \cdot 10^{-8} \text{ m}^2/\text{s}$$

$$R_p = 5.8 \cdot 10^{-5} \text{m} \qquad \rho_p = 725 \text{ kg/m}^3 \qquad \rho_f = 0.16 \text{ kg/m}^3$$

Mass transfer coefficient $k_p a$ — Because of the extremely small value of particle radius R_p, the estimates given in Table 8.1

have to be adjusted for size using the Glueckauf relations, Equation (8.6). We obtain

$$k_p a = 15 \frac{D_p}{R_p^2} = 15 \cdot \frac{2.8 \cdot 10^{-8}}{\left(5.8 \cdot 10^{-5}\right)^2} = 125 \ s^{-1}$$

This value is ten times higher than the maximum value for $k_p a$ listed in Table 8.2, which was based on a particle radius of approximately 1 mm.

Separation factor r — We wish to avoid a detailed calculation of the q^+ and q^- characteristics and hence propose using a "typical" value of r = 0.5 (favorable). The use of an estimate turns out to be justified since the operation takes place in a regime of high dimensionless distance N (see below) which renders the result relatively insensitive to the precise value of r.

Operating lines — The construction of the operating diagram proceeds as follows: a line is drawn from the feed point F(2) ($q_{F2} = 5.4 \cdot 10^{-2}$, $y_{F2} = 0.125$) to the origin, and another one parallel to it through F(1). The latter intersects the pure component isotherm (1) at the plateau value $q_p = 2.5 \cdot 10^{-2}$ kg/kg, $Y_p = 0.18$ kg/kg. The slopes of the relevant operating lines then evolve as follows:

For the leading front: $\quad \left(\dfrac{\Delta q}{\Delta Y}\right)_L = \dfrac{q_p}{Y_p} = \dfrac{2.5 \cdot 10^{-2}}{0.18} = 0.14$

For the rear zone: $\quad \left(\dfrac{\Delta q}{\Delta Y}\right)_R = \dfrac{q_{F2}}{Y_{F2}} = \dfrac{5.4 \cdot 10^{-2}}{0.125} = 0.43$

Dimensionless distance N — We have from Table 8.2:

$$N = k_p a \cdot \frac{z}{v}$$

which is multiplied by the correction factor 2/1+r. We obtain

$$N = 125 \cdot \frac{0.02}{0.0265} \frac{2}{1+0.5} = 126$$

Dimensionless time T — These are read from the fluid phase charts, Figures 8.2 to 8.6 for N = 126, yielding the following values

$\Delta Y / \Delta Y_o$	0.01	0.1	0.5	0.9	0.99
T	100	105	110	115	120

Breakthrough times — We refer to Table 8.2 for the expression for T:

$$T = k_p a \, t \frac{\rho_f}{\rho_b} \cdot \frac{1}{\Delta q / \Delta Y} \left(\frac{2}{r+1} \right)$$

which becomes

$$t = \frac{T}{k_p a} \cdot \frac{\rho_b}{\rho_f} \cdot \Delta q / \Delta Y \frac{r+1}{2}$$

For example, at the level of $\Delta Y / \Delta Y_o = 0.1$, we have for the leading front

$$t = \frac{105}{111} \cdot \frac{725}{0.16} \cdot 0.14 \cdot \frac{1.5}{2} = 452 \text{ s}$$

Additional results and a comparison with experiment are given below.

| $\Delta Y/\Delta Y_o$ | t_{ads} Leading Front | | t_{ads} Rear Front | |
	Predicted	Experimental	Predicted	Experimental
0.1	452 s	400 s	1210 s	1050 s
0.5	473	420	1270	1150
0.9	495	450	1380	1250

Illustration 8.7: Drying of an Organic Solvent

Organic solvents often have to be dried to extremely low moisture contents to avoid undesirable reactions with moisture-sensitive reagents or catalysts. One means of achieving this is by deep-drying ("super-drying") with zeolitic desiccants. Isotherms are usually highly favorable (see corresponding case of moisture uptake from gases, Figure 1.3) with separation factors approaching values of r = 0, but rising as the solvent becomes more water miscible. For the lower alcohols, the r value for zeolitic drying is of the order r = 0.1–0.5.[9] Paralleling this trend is a drop in diffusivity as the solvent becomes more hydrophilic, in line with the oft-made observation that water-soluble solvents are "difficult to dry". This has been attributed to the high activation energy required to separate the water molecules from its polar solvent neighbors. These parallel trends are revealed in the tabulations below[9]:

Substance	Water Solubility, ppm	$Dp \cdot 10^5$ cm^2/s
Carbon tetrachloride	84	12.5–14
Freon 12	150	—
Chlorobenzene	300	—
Xylene	440	2.5–5.8
Chloroform	970	1.3–2.7
Methylene chloride	1400	1.4–2.5
Freon 22	1800	—
Diethyl ether	12,600	0.23–0.31
Ethanol	∞	0.10
Acetone	∞	0.24–0.32

It will be noted that diffusivities vary by approximately two orders of magnitude between the limits of hydrophobic and water-soluble solvents. This does not in any way alter the equations pertinent to adsorption operations, provided due note is taken of the change in diffusivities. Regeneration of the bed is a much more complex affair which has received little attention so far as it involves evaporation of residual solvent as well as desorption of the desiccant itself.

We consider the drying of water saturated with carbon tetrachloride and diethyl ether representative of hydrophobic and hydrophilic solvents, respectively.

Operating lines and minimum bed requirements W_m — Zeolite equilibrium loading is near the sorbent saturation level of ~0.2 kg H_2O/kg solid (see Figure 1.3), so we can write

$$CCL_4 \qquad \frac{\Delta q}{\Delta Y} = \frac{0.2}{84 \cdot 10^{-6}} = 2.4 \cdot 10^3 \qquad W_m = \frac{1}{\Delta q / \Delta Y} = 4.2 \cdot 10^{-4} \frac{kg \ bed}{kg \ solvent}$$

$$C_2H_5O \qquad \frac{\Delta q}{\Delta Y} = \frac{0.2}{12600 \cdot 10^{-6}} = 16.7 \qquad W_m = \frac{1}{\Delta q / \Delta Y} = 6.0 \cdot 10^{-2} \frac{kg \ bed}{kg \ solvent}$$

Equilibrium breakthrough times: We set $\rho_b/\rho_f = 1$, $v = 0.01$ m/s, and $L = 1$ m. Using the usual design equations for equilibrium breakthrough, we obtain

$$CCl_4 \qquad t = \frac{\rho_b}{\rho_f} \cdot \frac{\Delta q / \Delta Y}{v} L = 1 \cdot \frac{2.4 \cdot 10^3 s}{10^{-2}} \cdot 1 = 2.4 \cdot 10^5 s$$

$$C_2H_5O \qquad t = \frac{\rho_b}{\rho_f} \cdot \frac{\Delta q / \Delta Y}{v} L = 1 \cdot \frac{16.7}{10^{-2}} \cdot 1 = 1.7 \cdot 10^3 s$$

Thus, both bed requirements and breakthrough times seem in order, at least under equilibrium conditions. The situation changes, however, under nonequilibrium conditions.

Mass transfer coefficients: We assume a particle diameter of 1 mm ($R_p = 0.005$ m) and compute mass transfer coefficients k_pa using the Glueckauf relation Equation 8.6.

$$CCl_4 \qquad k_pa = 15\frac{Dp}{R_p^2} = 15 \cdot \frac{13.2 \cdot 10^{-5}}{0.05^2} = 0.79$$

$$C_2H_5O \qquad k_pa = 15\frac{Dp}{R_p^2} = 15 \cdot \frac{0.27 \cdot 10^{-5}}{0.05^2} = 0.016$$

Thus, the mass transfer coefficients for carbon tetrachloride and diethyl ether drying differ by almost two orders of magnitude, due to the difference in diffusivities.

Dimensionless distance N and time T: N is calculated from the relation given in Table 8.2 and T read from Figure 8.2 for $\Delta Y/\Delta Y_o$ = 0.01 fractional breakthrough.

$$CCl_4 \quad N = k_p a \cdot \frac{z}{v} f(r) = 0.79 \cdot \frac{1}{0.01} \cdot \frac{2}{1+0} = 160$$

$$C_2H_5O \quad N = k_p a \cdot \frac{z}{v} f(r) = 0.016 \cdot \frac{1}{0.01} \cdot \frac{2}{1+0} = 3.2$$

$$CCl_4 \quad T = 150 \text{ read from Figure 8.2}$$

$$C_2H_5O \quad T < 0 \text{ read from Figure 8.2}$$

Breakthrough time: We consider the fractional level $\Delta Y / \Delta Y_o$ = 0.01 and use the correction factors of Table 8.3.

$$CCl_4 \quad t_a = \frac{T}{N} t_e = \frac{150}{160} \cdot 2.4 \cdot 10^5 = 2.3 \cdot 10^5 \text{s}$$

$$C_2H_5O \quad t_a = \frac{T}{N} t_e = < 0$$

Thus, CCl_4 breaks through after several comfortable hours; ethyl ether C_2H_5O, on the other hand, experiences premature breakthrough because of its low diffusivity. To remedy this, the fluid velocity has to be lowered to, say $v = 10^{-3}$ m/s. We then obtain

$$N = 32 \quad T = 26 \quad t_a = \frac{26}{32} \cdot 1.7 \cdot 10^3 = 1.4 \cdot 10^3 \text{s}$$

Thus, breakthrough is under 1 h but still comfortable.

Bed requirements: These are again established from the correction factors listed in Table 8.3. We assume velocity for C_2H_5O has been lowered to prevent premature breakthrough. Thus

$$CCl_4 \quad W_a = \frac{N}{T} W_m = \frac{160}{150} \cdot 4.2 \cdot 10^{-4} = 4.5 \cdot 10^{-4} \frac{\text{kg bed}}{\text{kg solvent}}$$

$$C_2H_5O \quad W_a = \frac{N}{T} W_m = \frac{32}{26} \cdot 6 \cdot 10^{-2} = 7.4 \cdot 10^{-2} \frac{\text{kg bed}}{\text{kg solvent}}$$

NOMENCLATURE

A	Area, m^2
a	Langmuir Henry constant
b_1, b_2	Langmuir isotherm constants, dimensionless
C	Concentration, arbitrary units
C_{pb}	Carrier gas heat capacity, kJ/kg K
C_{ps}	Sorbent heat capacity, kJ/kg K
D_p	Pore diffusivity, m^2/s
D_s	Solid phase diffusivity, m^2/s
f(r)	$= 2(r+1)$ for $0 \leq r \leq 1$, $= r^{-1/2}$ for $r > 1$
G	Mass flow rate, kg/s
G_a	Actual purge requirement, kg purge/kg bed
G_b	Mass velocity, kg/m^2s
G_m	Minimum purge requirement, kg purge/kg bed
H	Henry constant, kg carrier/kg sorbent
H'	Henry constant, mmol/g Pa
ΔH	Heat of adsorption, kJ/kg
$k_p a$	Pore mass transfer coefficient, s^{-1}
$k_s a$	Solid phase mass transfer coefficient, s^{-1}
K_{ow}	Octanol-water partition coefficient, m^3/m^3
K_{oc}	Water-organic carbon partition coefficient, m^3/m^3
L	Length of column or bed, m
LMTZ	Length of mass transfer zone, m
LUB	Length of unused bed, m
M	Molar mass
N	Dimensionless length (Table 8.2)
P_{tot}	Total pressure, Pa
q^+, q^-	Partial isotherms or characteristics
q_{ij}	$\dfrac{\partial q_i}{\partial Y_y}$
q	Sorbent loading, kg solute/kg sorbent
$\Delta q/\Delta Y$	Slope of operating line, kg carrier/kg sorbent
dq/dY	Slope of equilibrium curve, kg carrier/kg sorbent

Q	Volumetric flow rate, m^3/s
r	Separation factor (Figure 8.1), dimensionless
R	Recovery. Relative purge consumption, dimensionless
R_p	Particle radius, m
t	Time, s
t_a	Actual breakthrough time, s
t_e	Equilibrium breakthrough time, s
T	Dimensionless time (Table 8.2). Temperature, K
v	Fluid velocity, m/s
V	Propagation velocity, m/s
W_a	Actual sorbent requirement, kg sorbent/kg carrier
W_m	Minimum sorbent requirement, kg sorbent/kg carrier
Y	Fluid phase concentration, kg solute/kg carrier
z	Distance from inlet, m
ρ_b	Bed density, kg/m^3
ρ_f	Fluid density, kg/m^3

BIBLIOGRAPHY

The author was much aided in his deliberations and the preparation of his text by the following definitive texts and reviews:

1. Perry, J.H., Ed., *Chemical Engineers' Handbook*, Chapter 16, McGraw-Hill, New York, 1963.
2. Ruthven, D.M., *Principles of Adsorption and Adsorption Processes*, John Wiley & Sons, New York, 1984.
3. Yang, R.T., *Gas Separation by Adsorption Processes*, Butterworths, London, 1987.
4. Valenzuela, D.P. and Myers, A.L., *Adsorption Equilibrium Data Handbook*, Prentice-Hall, Englewood Cliffs, NJ, 1989.

Much of what is said about Henry's constants comes from the monographs of Ruthven and Valenzuela and Myers. The chapter in Perry's *Handbook* led us to the charts of Hiester and Vermeulen, which were later converted and extended into the form seen in this book. Yang's review of adiabatic sorption enhanced the author's own understanding of the topic. We did not enter into the intricate topic of PSA, which is well covered in the monograph:

5. Ruthven, D.M., Farooq, S. and Knaebel, K.S., *Pressure Swing Adsorption*, V.C.H. Publishers, Weinheim, 1994.

and to a lesser extent in the text by Yang cited above.

Much of what is said about equilibrium theory is based on previous publications of the author, notably:

6. Pan, C.Y. and Basmadjian, D., An analysis of adiabatic sorption of single solutes in fixed beds: pure thermal wave formation and its practical implications, *Chem. Eng. Sci.*, 25, 1653–1664, 1970.
7. Pan, C.Y. and Basmadjian, D., An analysis of adiabatic sorption of single solutes in fixed beds: equilibrium theory, *Chem. Eng. Sci.*, 26, 45–57, 1971.

8. Basmadjian, D. and Coroyannakis, P., Equilibrium theory revisited: isothermal fixed-bed sorption of binary systems. I. Solutes obeying the binary Langmuir isotherm, *Chem. Eng. Sci.*, 42, 1723–1735, 1987.
9. Basmadjian, D., The adsorption drying of gases and liquids, chapter 8 in: *Advances in Drying*, Vol. 3, Taylor & Francis, Washington, D.C., 1984.

Modified Hiester-Vermeulen charts are found in the latter.

Various conversations with D. Mackay (1995) revealed to us for the first time the approach taken by environmental scientists in addressing the problem of sorption of pollutants onto soil. This dedicated group of workers have developed their own methods of analysis of sorption equilibria which are well worth the scrutiny of those steeped in conventional methods (the author included). A very comprehensive summary of the data is given in works by Mackay and coworkers:

10. Mackay, D., Shiu, W.Y., and Ma, K.C., *Illustrated Handbook of Physical-Chemical Properties and Environmental Fate for Organic Chemicals*, Vol. 1, Monoaromatic Hydrocarbons, Chlorobenzenes and PCBs, Lewis Publishers, 1992 (and subsequent volumes).

We used these sources for the derivation of Henry's constants for water-soil systems, shown in Table 7.1.

We have reached into the general literature for a number of items. Saturation sorption capacities based on the theory of volume filling are taken from:

11. Grant, R.J., Manes, M., and Smith, S.B., Adsorption of normal paraffins and sulfur compounds on activated carbon, *AIChE J.*, 8, 403–406, 1962.
12. Grant, R.J. and Manes, M., Correlation of some gas adsorption data extending to low pressure and supercritical temperatures, *I&EC Fundamentals*, 3, 221–224, 1964.

Comparisons with experimental breakthrough curves given in Illustration 8.6 are based on the data of

13. Shen, J. and Smith, J.M., Adsorption rates for the benzene-n-hexane system on silica gel, *I&EC Fundamentals*, 7, 1, 1968.

FIGURE TITLES

FIGURE 1.1 Typical concentration front obtained during adsorption.

FIGURE 1.2 The six types of equilibrium isotherms encountered in sorption operations.

FIGURE 1.3 Moisture uptake on various commercial sorbents.

FIGURE 2.1 Propagation of discontinuous and expanding fronts along a type I isotherm.

FIGURE 2.2 Operating diagrams and concentration profiles for adsorption along a type I (Langmuir) isotherm.

FIGURE 2.3 Desorption from a type I (Langmuir) isotherm.

FIGURE 2.4 Adsorption along a type III (unfavorable) isotherm.

FIGURE 2.5 Desorption from a type III (unfavorable) isotherm.

FIGURE 2.6 Adsorption along an inflecting type II isotherm.

FIGURE 2.7 Desorption along an inflecting type II isotherm.

FIGURE 3.1 Characteristics (pathways) for binary Langmuir sorption.

FIGURE 3.2 Operating diagrams and profiles for binary Langmuir adsorption and desorption.

FIGURE 3.3 Graphical constructions for various types of binary Langmuir sorption.

FIGURE 3.4   The watershed point W.

FIGURE 3.5   The effect of feed location and isotherm shape on the enrichment of a light component displaced by a heavy component.

FIGURE 3.6   Displacement of solute from a type III isotherm by a type I isotherm.

FIGURE 4.1   Temperature effects in sorption.

FIGURE 4.2   Characteristics (pathways) in adiabatic sorption.

FIGURE 4.3   Desorption breakthrough curves at different purge temperatures.

FIGURE 4.4   Adiabatic adsorption on various commercial desiccants.

FIGURE 4.5   Adiabatic regeneration by hot gas purge of various commercial desiccants.

FIGURE 4.6   Minimum bed requirements in the drying of air with various desiccants.

FIGURE 4.7   Minimum purge requirements in the regeneration of desiccants.

FIGURE 5.1   Profiles and breakthrough curves for ternary Langmuir systems.

FIGURE 6.1   Propagation of injected solutes with linear isotherms.

FIGURE 7.1   Henry constants of pure gases on activated carbon and 5 Å zeolite.

FIGURE 7.2   Saturation capacities of gaseous solutes as a function of solute liquid densities.

FIGURE 8.1   Non-dimensionalized equilibrium isotherms.

FIGURE 8.2   Chart relating fractional fluid phase concentration = 0.01 to non-dimensionalized distance N, time T, and equilibrium uptake r.

FIGURE 8.3   Chart relating fractional fluid phase concentration = 0.1 to non-dimensionalized distance N, time T, and equilibrium uptake r.

FIGURE 8.4   Chart relating fractional fluid phase concentration = 0.5 to non-dimensionalized distance N, time T, and equilibrium uptake r.

FIGURE 8.5   Chart relating fractional fluid phase concentration = 0.9 to non-dimensionalized distance N, time T, and equilibrium uptake r

FIGURE 8.6   Chart relating fractional fluid phase concentration = 0.99 to non-dimensionalized distance N, time T, and equilibrium uptake r.

FIGURE 8.7   Chart relating fractional fluid phase concentration = 0.01 and 0.1 to non-dimensionalized distance N, time T, and equilibrium uptake r.

FIGURE 8.8   Chart relating fractional fluid phase concentration = 0.9 and 0.99 to non-dimensionalized distance N, time T, and equilibrium uptake r.

FIGURE 8.9   Chart relating fractional solid phase concentration = 0.01 and 0.99 at low values of non-dimensionalized distance N, time T, and equilibrium uptake r.

INDEX

Adiabatic sorption, 53
   adsorption, 60
   combined wave front (CWF), 57
   criteria for PTF/CWF formulation, 50
   design equation, 62
   desorption, 60
   maximum enrichment, 63
   minimum bed requirement, 62
   minimum purge requirement, 62
   plateau, 61
   pure thermal wave front (PTF), 55
   q-Y (operating) diagram, 60
   regeneration, 69
     See also Desorption
   steam regeneration, 63
   temperature rise, 54
Adsorbents, commercial properties of, 5
Adsorber operation, 3
   range of parameters, 6
Adsorption breakthrough curve, 2
   profile, 2
Adsorption Equilibrium, 7
   Henry constants, 87
   saturation capacities, 92

Binary Langmuir sorption, 33
   adsorption, 37
   desorption, 37
   mutual displacement, 37, 41
   on clean bed, 39
   on preloaded bed, 39
   prediction of breakthrough, 123
Breakthrough time, 15

Characteristics, 33
Clearance of river bed, 120
    of contaminated soil, 24
Creative doodling, 37

Design equation, 15
Desorption breakthrough curve, 14
Discontinuous front, 10
Drying
    air drying operations, 65
    minimum bed requirements for drying of air, 72
    minimum purge requirements for drying of air, 73
    of a carbon bed, 29
    of gases, 6
    of an organic solvent, 123

Effective equilibrium curve, 62
Expanding front, 10

Henry constants
    estimation of, 93
    for activated carbon, 90
    for 5A zeolites, 90
    for soil-water systems, 89
HTU
    See Sorption with transport resistance

Isotherms
    favorable, 4
    inflecting, 4
    Langmuir, 7
    linear, 4
    of moisture, 5
    types I-IV, 4
    unfavorable, 4

Linear chromatography, 81
    general features, 81
    propagation of rectangular pulse, 83
    minimum column length, 84

LMTZ
    See Sorption with transport resistance

Mass balance, 9
Minimum bed requirement, 15
    bed volume, 119
    purge requirement, 15
Multicomponent sorption, 75
    linear isothermal systems, 76
    nonlinear isothermal systems, 77
    purge of nonlinear solutes, 78

NTU
    See Sorption with transport resistance

Operating diagram
    Type I, adsorption, 17
    Type II, desorption, 18
    Type III, adsorption, 21
    Type III, desorption, 22
    Type II, adsorption, 27
    Type II, desorption, 28
Operating line, 16

Removal
    of aqueous pollutants from groundwater by soil, 23
    of aqueous pollutants by activated carbon, 76, 119
    of butane from N_2-N_2 by activated carbon, 76, 119
    of methane from air by activated carbon, 112
    of solvent vapor from air by activated carbon, 25
Rule 1 (IF Rule), 26
Rule 2, 36

Sorption with transport resistance
    design charts
        fluid phase, 102
        solid phase, 107
    Glueckauf relation, 100
    grafting onto equilibrium effects, 110
    height of transfer unit (HTU), 96
    length of mass transfer zone (LTMZ), 96

length of unused bed (LUB), 96
mass transfer coefficient from measured breakthrough data,
    118
number of transfer units (NTU), 96
separation factor, 98
sharp sorption fronts, attainment of, 111
transport coefficients, 99
Specific consumption
    bed, 12
    purge, 12

Velocity
    of mass, 3, 11
    of propagation, 3, 11
    superficial, 3, 11